ANIMALS ON THE FARM

ANIMALS on the FARM

Their history from the earliest times to the present day

Judy Urquhart

MACDONALD & CO
LONDON & SYDNEY

First published in Great Britain in 1983 by
Macdonald & Co (Publishers) Ltd
London & Sydney

Maxwell House
74 Worship Street
London EC2A 2EN

ISBN 0 356 07897 3

Filmset, printed and bound in Great Britain by
Hazell Watson & Viney Ltd, Aylesbury, Bucks

Contents

Acknowledgements

The author and publisher wish to thank the following for permission to quote extracts from the works listed below:

William Collins & Sons Ltd, *Guide to Good Food and Wine: A Concise Encyclopaedia* by André Simon (1956)
Penguin Books Ltd., *The Canterbury Tales* by Chaucer (Revised Edition, 1977)
Laurence Pollinger and the Estate of Frieda Lawrence Ravagli, *The Complete Poems of D. H. Lawrence* by D. H. Lawrence published by William Heinemann Ltd.

List of Illustrations

Sources of the Illustrations

BPCC/Aldus Archive; 48
Country Life Archive, National Museum of Antiquities of Scotland; 49
Courtesy of Christie, Manson & Woods Ltd.; 13
Farmer's Weekly; 2, 8, 14, 15, 16, 19, 20, 29, 30, 31, 38, 39, 51
Michael Holford; 21, 35
Imperial War Museum; 43
Courtesy S. W. Kenyon, Wellington, Somerset; 33
The Mansell Collection; 6, 22, 24, 25, 26, 27, 34, 36, 40, 44, 47, 50
The National Gallery of Scotland, Edinburgh; 3
The Rare Breeds Survival Trust; 12
John Rea Studios, Shrewsbury; 19
Rothamstead Experimental Station; 9
The Trustees of Sir John Soane's Museum; 45
John K. Wilkie, Edinburgh; 20
Royal Library, Windsor. Reproduced by gracious permission of Her Majesty Queen Elizabeth II; 4
University of Reading, Institute of Agricultural History and Museum of English Rural Life; 5 (Graphic Photo Union), 7, 10, 17, 18, 23, 28, 33, 37, 41, 42
Judy Urquhart; 46
The Zoological Society of London; 1, 11, 32

Introduction

Farm animals have shared and shaped our history making possible our civilization. They have helped form our landscape and allowed most human achievement. The traffic between man and farm animal has always been two ways. Although man has long held the ascendancy because of his greater intelligence and articulateness, animals have, while accepting man's protection, provided the mainstay of his life. Thousands of years ago, as soon as farming began, they released people from time-consuming chores and gave them the time to develop further skills. Oxen hauled huge weights, dragged ploughs and raised water from wells. Horses brought unprecedented speed to warfare and communications. Sheep provided an annual crop of wool while alive and, at their death, left a legacy, as did cattle, pigs and goats, of meat, skin and bone. Poultry provided eggs and meat and feathers for beds and writing. They have fed us, clothed us, helped us build our houses, temples and pyramids.

Seeking to save himself more time, and to make farm animals even more useful, man began breeding them selectively. Changes that occurred naturally and others, determined by environment, revealed variants and these man accentuated, the range reflecting his multifarious needs and differing tastes. Animals were bred for lean meat, fat meat, more meat, power of traction, resistance to disease, drought and flood. They were bred for optimum combinations of various qualities, for different climates, soils and cultures. Sometimes they were bred simply to look handsome. The result is a range which goes from the mighty Shire horse to the minuscule Shetland pony; from the faun-like Soay sheep to the large, lanky Leicester; and from the wild White Park cow to the docile Jersey.

But who knows much about the background of these essential animals besides a few farmers? In earlier times all men's lives were bound up with their stock, they lived close to them and their care was part of daily existence. This symbiotic relationship no longer exists nor is it possible. Today the average person lives in a town where he cannot keep such animals, and even in the country the housing of animals by intensive farmers means that they are not seen in fields in the same numbers. Such factors have estranged modern man from his former close relationship with farm animals.

This book is an attempt to trace the history of that relationship. The story began 11,000 years ago in the Middle East and differed with each species of animal as, it is hoped, the following chapters will reveal. These are set in the sequence of the historical importance of the individual animals. At every stage a different one was vital. The goat, browsing on the all encompassing forests and helping with their clearance, was the main support of Neolithic man. Trees remained the predominant feature of the

1

landscape in the early middle ages and the pig, which could sustain itself and man from the forest floor, prevailed. By Tudor times large areas of forest had been replaced by grass and the sheep changed this into wool and made fortunes for its owners. In the 19th century the horse took the cow's position at the plough, releasing it to provide meat and milk to feed the urban populations of the Industrial Revolution. Today, a lack of space and money mean that intensively-housed cheap poultry provide most of the world's meat. Finally a chapter on Conservation is included, a subject which cannot be ignored in the face of man's increasing ability to manipulate the world with his scientific and technical expertise.

Why out of all the thousands of species of animals that inhabit the earth have so few been domesticated? And why, in Britain, have we only chosen to bring five animals, the goat, pig, sheep, cow and horse, and four birds, the hen, goose, duck and turkey, under our aegis? The reason is that it takes special qualities to fit animals for domestication. First they must be strong enough to withstand removal from their mother at an early age, be easy to look after and breed freely in captivity. Secondly, they must have an instinctive liking for man, enjoy comfort and confinement and control. Thirdly, they must be useful, usually as a source of food but often also in such ways as by providing wool for clothes, fat for light, and transport or traction.

Animals suited to domestication only emerged with the spread of grass. As the first Ice Age retreated it revealed a tundra-like country which became covered in woods of pine and birch, followed by elm, oak and hazel. Then, as the rainfall decreased, the forest diminished, low-growing plants and grasses appeared and with them grazing animals developed. Forty million years ago in the Upper Miocene Age, sheep and antelope appeared on the scene and later, in the Pliocene Age, cows, goats and horses. Not until 9000 BC did man develop the wit to take animals into captivity. There was wild game in abundance, so what prompted the idea? It could have evolved naturally from his existence as nomadic hunter. From automatically following the migration of his prey, in much the same way as, more recently, Laplanders went with the reindeer and American Indians the bison, he gradually began to assert an influence over their movements. Alternatively domestication could have begun from the keeping of young animals as pets. Wolves and jackals on the scrounge round camps were slowly absorbed into the community, as pilfering turned to feeding and being fed, and with it the wolf changed into a dog.

Pigs are scroungers like wolves and their domestication probably happened in a similar way. Sheep, cows, goats and horses move in flocks and herds and it may be that initially the young of these larger animals were captured and reared but killed before they reached their full size and became impossible to control. The next stage could have been for the more docile females to be kept for their full life span and replacements bred by tethering a female on the outskirts of a village until mated by a wild male.

As the animals were tamed their looks changed. Most became smaller, partly because man could not cope with large animals and chose small ones and partly because he could not feed them as effectively as they fed themselves in the wild. Then they developed more colour. Most wild animals give the impression of being a tawny brown but actually their coats are made up of many different coloured hairs and with domestication one colour came to dominate over another. Their bone-structure altered and the face, especially of pigs and sheep, became shorter while the jaw muscles contracted as they were no longer needed to masticate tough herbage. The other muscles in the body also shrank as their power was no longer required to flee from enemies and cover long distances in search of food. Once the animal no longer has to fend for itself the brain shrinks and the senses used to alert it to danger, such as sight, hearing and scent, become less acute. In fact tame animals, molly-coddled by man, fail to grow up. None of these alterations are permanent as animals set back in the wild revert within a few generations to their former wild shape and character. This has been seen with mustangs in America and pigs in New Zealand.

The atrophied characteristics produced by domestication have been further exaggerated by man's selective breeding. The process has taken an incredibly long time. It took six thousand years for Neolithic man to move from the Middle East to Europe and then about five hundred more for him to reach Britain. The people left behind acquired arts that resulted in the Egyptian, Greek and Roman civilizations but those pushing through the virgin forest, fanning out east and west, making occasional settlements where they practised a crude kind of farming on their semi-wild sheep, goats, cattle and pigs, remained in a primitive state.

Around 2500 BC some of these small, dark Neolithic farmers had reached the north coast of France. Pressurized by warring tribes coming up from the south they looked for new land and decided to cross the sea dividing them from the cliffs of England. As the channel was narrower in those days it would not have been such a formidable task despite the fact that Neolithic man's means of transport was a coracle-like boat.

His arrival coincided with a change in the climate. Britain was becoming drier and, deprived of water, the trees died away on the southern chalk and oolitic limestone hills. Neolithic man found clear space on the tops of these hills in which to graze his animals and establish settlements and, in a triangle formed by Devon, Sussex and Bedfordshire, left remains of his great camps or kraals. The climate continued to dry and aided by this, the ravages of his livestock and his polished flints which, it is said, could fell a fir tree seventeen centimetres in diameter in five minutes, Neolithic man gradually destroyed more trees. In the clearings he made more settlements and lived in round or oblong huts with underground storage pits but his stock wandered free and fed themselves on the indigenous vegetation.

Neolithic man was followed across the Channel by people who knew about metal and they introduced the Bronze Age to Britain and more efficient cutting tools. Then, in about 500 BC the first of successive waves of Celts arrived from Europe. Taller, fairer, more muscular, and more civilized,

3

these people brought in a new kind of cow and perhaps a different sheep. They constructed immense earthworks, enclosing areas of anything from six to eighty acres into which they herded their stock, and built houses on the downs, but in the hills and valleys they began a system of farming now called 'transhumance'. By this method communities and stock live safely in the valley bottoms during winter and move up to summer hill farms in spring. A second body of Celts came between 150–200 BC. An artistic people, they flourished mainly in the west of the country and built the marsh village of Glastonbury. A third set of Celts arrived in about 75 BC. The teutonic blooded Belgae probably brought the first domesticated horses, hens and geese as well as new methods of agriculture. They lived mainly in the south and south-east of England, made settlements in open places by river crossings, cleared extensive tracts of forest with their superior implements and kept their stock in little fenced fields. The Belgae's greatest contribution to agriculture was a heavy wheeled plough drawn by a team of oxen. This turned the sods of England enabling crops such as beans, pease, millet, oats and flax to be grown.

These three cultures existed side by side until, in the 1st century AD, most of Britain was conquered by the Romans. Outside the areas of Roman influence, in the far western parts of England and Wales and in Scotland, the Celts continued with their former existence and kept to their old ways of farming. However, in the parts under Roman rule life changed dramatically. In the thousand years before the Romans arrived the climate of Britain became wetter again and the land was full of morass and bog. The Romans drained huge areas and reclaimed low-lying districts by building enormous embankments. They also cleared great tracts of forest, especially on Salisbury Plain, Cranborne Chase and the South Downs and brought land into cultivation. Corn was planted and Britain became one of the most important grain-producing countries in Europe. And the Romans probably introduced their own types of sheep, cattle, horses and poultry.

In the upland and open areas the animals were herded and put into entrenchments at night as wolves were plentiful but, on the the villa farms of the south and south-east, they were kept in much the same way as today. Remains found at settlements in Cranborne Chase reveal that sheep and oxen were the most favoured foods, horses and pigs came second and rations were supplemented by the occasional deer, fowl and dog. Stock was moved along the newly constructed roads to supply Roman legions guarding the outposts in the north and west with food, clothing and tent hides.

The order created by the Romans fell into disarray when they began to withdraw early in the 5th century, but from the 5th to the 7th century Britain was attacked by a succession of tall, blond, blue-eyed teutons. These Angles, Saxons, Jutes and Friesians left their homes after planting their crops in spring, returning for the harvest and to prepare for winter. Gradually they started settling the areas they raided round the coast and began to move inland and impose their tribal way of life and agriculture on the people they subdued, congregating in 'hams' or 'tuns' under chiefs who

4

assigned each family a 'hide' or land. A hide averaged 120 acres but varied according to the nature of the soil as it was measured by the amount of arable land a man with a full team of eight oxen could plough. The Scandinavians introduced their own red cows and probably a kind of black-faced sheep and kept their animals in stalls at night but drove them out to graze during the day. In the 8th century the Norse turned their attention to Scotland, made permanent settlements on the mainland and on many of the outer islands, including Shetland, and brought in more animals.

The Norsemen held onto their northern territories but the kingdoms of the south were overtaken in the 11th century by the Normans. 'That dark and ruthless people, who combined the stern qualities of their Norse origin with a superficial polish of French civilization' stabilized rather than changed the farming system the Saxons had established. Their only contribution to the farmyard was the duck but the Domesday Book, their Great Survey of 1086, contributed much to a knowledge of the current state of agriculture. It shows that sheep were kept on open areas of country, cattle were raised in woodland and upland pastures and half-wild herds of pig of indeterminate breed ranged the forests. Some dairy farming was practised but it was based on sheep and goats not cows. Cattle were primarily used as beasts of traction and horses carried knights into battle. The numbers of pigs exceeded that of any other animal.

In the 14th century people grew their produce for their own consumption and not for sale. Men, and the muscles they possessed, were more vital than money which was why the Black Death had such a devastating effect on agriculture. This plague reduced the meagre population of Britain by at least a third, if not a half, and affected livestock almost as severely. Sheep were particularly badly hit and even poultry and bees were not exempt.

The Black Death contributed to the extinction of the manorial system which tied men to the land and exerted strict control over their movements. The manorial system was already faltering and, with the devastation of population and stock, it fell rapidly into decay. In the subsequent chaos the lord of the manor could no longer maintain his authority and peasants demanded and were given wages or were allowed to become tenant farmers. Others left the country for the towns which were growing fast during the period and offered wages which amounted to more than a poor peasant could expect to gain from his small holding of diseased land and stock. The departure or death of so many peasant owners left vast areas of country vacant and this was repossessed by the manor and turned into sheep walks; for sheep required few men for their maintenance and wool was fetching high prices.

Farming organized on a money basis came into being, and a new class of yeoman farmer grew up, controlling his own land and hiring free but landless labourers. Typical of these prosperous yeomen farmers was Hugh Latimer. He farmed in Leicestershire at the end of the 15th century, rented about 200 acres of arable land with rights to common pasture, employed six men on the farm as well as women servants in the house and dairy, ran 100 sheep, milked 30 cows, kept oxen for ploughing and a horse for riding or fighting for the King.

The ruin of many noble families by the Wars of the Roses further contributed to the redistribution of land. And Henry Tudor's defeat of Richard III on Bosworth Field in 1485 marked the beginning of a new age and a revived spirit. The population was still pathetically small, probably not more than three million, but this 'grete lake of pepul and inhabytantys' was compensated for by a rise in livestock numbers. Only a quarter of the country was under crops; the rest, Harrison states, was 'wonderfully replenished with neat and all kinds of cattle' and 'Where are oxen commonly made more large of bone, horses more decent and pleasant in pace, kine more commodious for the pail, sheep more profitable for wool, swine more wholesome of flesh, and goats more gainful to their keepers than here with us in England?'

Money came into circulation allowing borrowing and, with it, expansion. Lord Ernle in *Farming Past and Present* (1912) says that 'Commerce permeated national life. Feudalism was dead or dying, trade was climbing on the throne. The Middle Ages was passing into modern times.' And they were passing on the back of the sheep. The money to be made from wool excited ruthless landlords to grab common and any other kinds of land for sheep pasture and to set exorbitant rents, driving small tenants from the land. There was some sense in the greedy actions of the 'cormorant' landlords. The land was exhausted and sick from centuries of producing crops unsustained by proper manure. The impracticality of strip farming and keeping stock all together on commons where disease flourished and breeding was uncontrolled was removed when land was enclosed and separately owned. But 'like all transition periods', commiserates Lord Ernle, 'it was full of suffering for those least able to adapt themselves to altered conditions.' Those who did adapt 'wexed verie Rich men' and were 'able to and do daily buy lands of unthrifty gentlemen and make . . . their sons gentlemen.' Wool was the chief source of their wealth and of the revenues of the Crown. The trade was mainly controlled by foreigners and a maze of monopolies and patents. Elizabeth I changed all this, she restricted exports and encouraged the home manufacture of cloth.

Towns grew in population and the needs of the 'great urban mouth' were fostered by the dissolution of the monasteries which released more land to the yeoman farmer and a new type of businessman farmer. Both these new men helped transform the cultivation of the soil and the management of livestock. The prices fetched by butter, cheese and meat rose to equal wool, changing the cow, in the 16th and 17th centuries, from being primarily a begetter of plough oxen to a giver of milk and meat, and the sheep from a wool to a mutton producer. A lack of forest grazing put the pig in a stye. Turkeys were introduced from South America and the poultry yard ceased to contain a range of birds almost rivalling the hedgerow in variety and assumed much its present form. The Dutch and Belgians supplied the knowledge to grow forage and fodder crops in fields and these foods enabled more livestock to be kept through the winter and allowed all to reach their full growth.

The Civil War checked these improvements. The dispute never affected the whole country but it did create a sense of insecurity that caused 'the promise of agricultural progress to perish in the bud.'

Agriculture languished, the demand for patents for new improvements dwindled and Pepys complained that 'Our gentry are grown ignorant in everything of good husbandry.' The most far-reaching effects of the Civil War were felt by the horse. Knights in armour were useless against the new gunpowder and the huge horses bred to carry their weight, no longer required by the army, found their way onto the farm. They were able to draw carts and ploughs with greater speed and efficiency than oxen and gradually took their place.

Life revived when peace was made, farmers took heart and their interest was stimulated by a mass of agricultural writers. Leonard Mascall and Gervase Markham wrote on every matter. Many exploited the new ideas, especially a fresh, enterprising breed of landlord who had bought the estates confiscated by the Commonwealth. In this regime only the agricultural labourer suffered. His wages were low and his traditional hold on the land was increasingly threatened by enclosures. Living in isolated villages with roads often impassable because of mud, the progress of the outside world rarely reached him and he had little incentive to raise more from the soil than would meet his immediate wants.

In the late 17th century Gregory King in his Statistical Accounts of England and Wales estimated that the total acreage was 42 million, of which 10 million was composed of heaths, moors, mountains and barren lands; 11 million were devoted to arable crops, 1 growing flax, hemp, saffron, woad and other dye plants and 10 growing corn, pease, beans and vetches; and 10 million was meadow pasture, land on which grazed 600,000 horses and asses, 2 million pigs, 4½ million cattle and 11 million sheep. King deduced that the population was 5½ million of whom 4,100,000 lived in the country and 1,400,000 in towns. The poorest counties were those of the north and the richest those lying immediately above the Thames. This distribution reflects the rising importance of London whose rapid growth was causing alarm lest 'the Head should become too big for the body' as its population had reached 530,000.

London's soaring population requiring continual quantities of fresh food set a new cycle in motion. Forage and fodder crops enabled livestock to survive the winter in good condition and to be kept in greater numbers. More stock meant more manure, raising more crops and more, larger animals but of limited quality. Until the tentative experiments of the 17th century were taken up and developed by such great farmers as Lord Townsend, whose enthusiasm for turnips earned him the nickname of Turnip Townsend, Jethro Tull who abandoned the law for farming and invented the seed drill, and a host of other livestock improvers, neither crops nor animals could show their full potential. Some modification had

7

already been made with the introduction of foreign blood, but it was Bakewell and his contemporaries who, with a foresight of markets to come, redesigned the local types of sheep, cow, pig and horse. The theories of breeding they advanced awakened the whole of Britain to the idea that certain characteristics could be selected in individual animals and these perpetuated and fixed in their progeny until, after several generations, the stock 'bred true' and a breed was established.

These breeders started a fashion for farming. In spite of the view expressed by Dr Edwards in 1783 that 'Gentlemen have no right to be farmers', everyone from Kings to peasants took part. George III began a model farm at Windsor and gained the title of Farmer George. The Duke of Bedford at Woburn, Lord Rockingham at Wentworth, Lord Egremont at Petworth and Thomas Coke at Holkham, among others, entered the field with zeal. The writings of agricultural journalists such as Arthur Young were eagerly perused for the latest advances, new implements were immediately tried and cattle shows, wool fairs and ploughing matches were held in various parts of the country.

More land was enclosed and brought into efficient production. Common lands were berated as being 'Seminaries of a lazy Thieving sort of People'. The protests of the peasants were ignored and those whose common lots of consolidated land were too small to support a family, without common rights of woodland for fuel and repairs and grazing for animals, moved to town. The magnet of town markets, and especially the London market, contributed to the growth of droving and animals were moved all over the country.

The progress of the pioneers marked the final end of mediaeval open-field farming and divorced the peasantry from the soil for ever. The changes were further accelerated by the Napoleonic Wars. During the war corn rose in price and food was scarce which resulted in an over-expansion of farming. When the war ended in 1815 this glut of produce combined with a glut of workers, made up of discharged servicemen and redundant industrial workers, led to a depression which lasted until 1836. Then the money farmers had made from the war, coupled with the agricultural advances made in the 18th century, began to produce results. Fields were spread with more and better fertilizers and stock was better fed and housed. General utility animals became a thing of the past and a host of new specialized breeds were developed. The value of feed was assessed and no longer was the stockman's eye the sole judge of the correlation between what an animal ate and what it produced. An interest in manufactured foods was excited by research done at the Moglin Institute in Germany and the enthusiasm of Robert Boutflour and James Mackintosh; grass was preserved as ensilage and rape was pressed into cattle cake.

But it was difficult to diffuse knowledge among people who could not afford books and were often illiterate and suspicious of novel methods. The average farmer needed some practical demonstration to convince him of the benefits of the innovatory methods which was provided when the Royal Agricultural Society was formed in 1838.

Farmers began to flourish as never before and the growth of wheat and other grain crops expanded. Then, during the second half of Queen Victoria's reign, colonists began to exploit fully the virgin soils and reliable climates of the Middle West of America and Canada, Australia and Russia, and this easily-won produce poured into Britain. The scales were further weighted against the British farmer by three disastrous seasons which culminated in a summer of icy rains in 1879. This sounded the death knell of Britain as a corn-growing country and had a devastating effect on livestock. From the 1880s prices of fat cattle and sheep declined in the face of cheap meat from America, Canada and then New Zealand and Argentina. The development of the refrigerated ship and store further increased the imports. Mutton rose from 181,000 cwts of boiled and tinned meat in 1822 to 3½ million cwts of frozen carcase in 1899.

Parliament was powerless to legislate against foreign imports because food was the currency used by these countries to pay for English manufactured goods. Ruin and bankruptcy were the fate of many farmers and in 1893 a Royal Commission was appointed to investigate the depression. It revealed that men of capital and energy on well equipped farms grew the best crops and livestock and, even on heavy land, had weathered the storm; that small owner occupiers who employed no labour had also fared well, that market and fruit farming were profitable and that milk made money. The Royal Commission instigated a series of acts designed to assist the farmer; the Improvement of Land Act gave landowners increased facilities for borrowing money to carry out improvements, and the Fertilizers and Feeding Stuffs Act and the Sale of Food and Drugs Act were designed to prevent the adulteration of cake, fertilizers and dairy produce.

But the basis of farming had changed for ever. Corn-crops were reduced and woodlands began to be treated as commercial forests. The development of the milk trade, dairying, pasture-farming, flower-growing, market-gardening and poultry-keeping are all products of this period. Attention was turned to breeding pedigree livestock whose ancestral lines were written in the herd books published by the new Breed Societies. Their breeding was further influenced by the science of hereditary genetics. This was first advocated by its inventor, an Austrian monk Gregor Mendel, in 1865 but took time to be accepted. From his researches into the hybridity of plants Mendel established laws of dominant and recessive characteristics. This new science opened doors to stock-breeders, providing the knowledge of blending the useful characteristics of different stock into one type and enabling new strains to be started.

The best of these new strains were exported to pass their genes on to progeny in the USA, Canada, Australia, New Zealand and Argentina. Their offspring returned as frozen carcases. Recognizing that they could never compete with all this cheap meat, British farmers reorganized themselves to changed circumstances. Realizing that the public still valued succulent tasting, fresh meat and that 'The best farming is that which pays the farmer best' they said 'God speed the plough on every soil but our own', laid land down to grass and brought cheap foreign feeds to give their stock in winter. The tillage area of England dwindled from 6¾ million to 3 million acres and the area devoted to stock grew to 36 million. This drove yet more people from the land as stock requires fewer men for its maintenance than tillage.

During the First World War the balance changed again. The blockade meant that all types of food had to be produced at home and any available labourers, chiefly women, were put to work on its growth. The government rationed food and controlled prices. At the end of the war there was a slump similar to that caused by the Napoleonic war; there was an over-production of food, prices fell and the labour market was flooded. In 1920 the Government introduced an Agriculture Act and laid the foundations of the state protection of farming.

Later the system of Free Trade was resumed and the markets were once again deluged with American grain, New Zealand and Argentine meat and Danish eggs and bacon. British farmers went back to concentrating on the two commodities least exposed to foreign competition, fresh meat and milk. Many servicemen spent their annuities on small holdings, specializing in market gardening and the production of poultry and veal.

Cheap imports mounted in the 1930s and put an unbearable strain on farmers. Thousands went bankrupt, land was neglected and the numbers of livestock fell. The Government stepped in with guaranteed prices for fatstock and milk and, in anticipation of another war, gave grants for crop growth. The Government's change of attitude marks a major divide in the history of agriculture as, from now on, farmers had security which was something they had never experienced before.

After the Second World War livestock became the most important element in British farming. However, crops and livestock are to an extent complementary even though their distribution is dictated by climate and soil and in 1970 80% of all agricultural land was used for the support of animals. 55% of these were cattle (two-thirds of which were dairy cows and one-third beef cattle), 23% were sheep, 7% were pigs, and 7% were poultry, while farm horses and goats were considered such a minor part of farming that their numbers were no longer recorded.

Many animals have disappeared from fields and into houses. The intensive farm is a product of expensive land and feed. Alterations in human diets have also boosted the factory farm. An increasing proportion of the food people buy is processed, pre-packed and oven-ready, and this standardization requires large scale production and quality control.

Today's agriculture is the nation's largest industry but farmers comprise only 3% of the population; they number 90,000 with another 360,000 working in related industries. The average farmer has an invested capital as large as many small factories. The amount of farm land is diminishing; thousands of acres are swallowed up annually by 'Rurban' fringe development and connurbations now cover 6 million acres. The rest is divided up between 29 million acres of arable, grassland and other cultivations, 16 million acres of hill and moor, and 5 million acres of wood and forest. However, since 1945 farmers have doubled the quantity of produce they extract from the land and now provide 70% of all the food consumed by a population that numbers over 55 million.

How long this situation will last is a matter for speculation. The cost of energy, fertilizers, chemicals and machinery continually rises making food ever more expensive. Escalating prices and the fear of heart disease and other ills associated with fatty foods are turning the public away from conventional sustenances such as meat, milk, eggs and glass-house crops to cheaper, fatless foods like fish and pulses. Dairy herds have decreased and there has been a fall in beef production. The most popular meats today are pork, lamb and poultry and, as a result, the numbers of these animals have increased. Prices for cereals have risen and with them the acres devoted to these crops which, in 1980, were at a record level.

Further changes could occur because of protests by animal welfare organizations and public horror at some of the methods used to grow factory produced meat. There are indications that public resistance is having an effect on poultry and pork sales. Organic farmers make fewer profits but they sell all they produce. However, these people are in a minority, the majority of British farmers set their sights on deriving the utmost from each ounce of soil for which they receive prices guaranteed by EEC regulations. Their expertise in livestock breeding, gleaned over the centuries, results in stock that is sought after all over the world. But live animals are heavy and expensive to transport and the sale of frozen embryo or even cloning may be the next step.

Animals are regarded as 'a department of the farm which must pay its way' and carcases have to conform to exact specifications. In America researchers are talking about making cows swallow transmitters so that computers can monitor their internal workings and farmers ascertain 'better how to fatten them'. They forecast that in the next twenty years 'We will be producing more meat less expensively, and we will have the opportunity for much more export.' Will this be the way forward or will energy shortages and western food surpluses lead to a lower production of more naturally grown animals? Perhaps we will end up buying cheaply-produced food from Saudi Arabia or concentrating on quite different stock like fish and deer.

The Goat

The goat is the animal that has most changed the Old World and made space for the new. Its methodical mastication of the forests of the Middle East enabled man to create clearings in which to cultivate crops and keep other animals. The Bezoar goat was the first farm animal. This goat with its long shaggy coat and sweeping, scimitar shaped horns roamed wild in the thickly forested mountain ranges which stretched from Sind in Pakistan across the Middle East to the Balkans. Around 8000 BC men in the Middle East took the Bezoar into captivity, choosing it in preference to other animals because it was able to survive on a rougher, scantier diet in a tougher climate and could provide meat and milk for sustenance, skins for clothing, water containers and boats, sinews for bows, and fat for light.

Sheep quickly followed the goat into domestication. Once there was grass for their grazing, their warming wool and more tender meat made them preferable. Firewood, however, was still needed for cooking fuel and, as the goat demolished all supplies in its immediate vicinity, man was forced to move in its wake. It was centuries before it occurred to man to replant the trees destroyed by the herds of goats and this short-sighted policy created the deserts of the Middle East and Sahara, and the ecology of the countries bordering the Mediterranean was drastically altered by the soil erosion that follows the extinction of the flora and fauna.

Goats were essential to man while he remained primitive and nomadic and went with him when he began migrating out of the Middle East in about 3000 BC. By this time, archaeologists reveal, man had developed forms of goats and some had gained a twist in their enormous horns. More variations were added when, as man travelled, he took into captivity the different kinds of wild goat he found in his path. In the Himalayas he domesticated the Markhor, the largest wild goat, which eventually developed

12

into the Cashmere and Cheghu breeds and, in the Mediterranean basin, a small, short-haired goat called *Capra pisces*. In time, crosses between Markhor and Bezoar produced the Angora and others between Bezoar and *C. pisces* the ancestors of most European goats.

All these early goats were magnificent looking creatures, standing about four feet at the shoulder, varying in colour from grey to black and brown with wild, hairy coats, vast curving horns and long ears.

Babylonian astrologers believed the Milky Way was a she-goat and that at the top of a sacred mountain stood Polaris, a sacred he-goat 'the highest of the flock of the night'. The Jews took a different attitude, considering the goat the lowest of the low, made it the 'scape-goat' and sacrificed it for the sins of others. The Bible talks of 'separating the sheep from the goats' but its writers did not spurn the goats' products, weaving their hair into cloth, drinking their milk, eating kids' meat and inflating their skins as bottles and boats. The Assyrians also used goats' skins as swimming bladders or buoys.

The ancient Egyptians had vast herds of goats and elevated them into Gods. The Greeks also placed a high premium on the goat, used all its products and worshipped the animal as a symbol of fertility. The Roman descriptions of goats were more practical; they divided them into two classes; those with fine hair whose horns were sawn off and those with shaggy hair whose horns were allowed to grow. A good he-goat, according to the agricultural writer Columnella, should have two small wattles hanging from its neck, a large body, thick legs, a full short neck and very long, flaccid, shiny hair. Pliny commends their intelligence.

The first domesticated goats to reach Europe arrived in Neolithic times with invaders from the south-east. Bones of scimitar-horned goats are frequently found in the early layers of Swiss Lake Village settlements. In later layers sheep bones replace goat bones which is probably another indication that, once civilization was established and the goat had cleared the indigenous forest, it was largely replaced by the more useful sheep.

The goat that arrived in Britain with Neolithic man in 2500 BC had long, blue-grey hair, short legs, and sweeping horns and for a while was his most important animal. Once again goats were called upon to chew a path through the forest, provide man with milk and meat for his sustenance and skin for his clothing. But after they had cleared the trees, brambles, briar, ivy, gorse, heather, thistles, nettles and docks from the hill tops and downs the way was open for more pastoral animals and the goat's services were dispensed with.

A small, short-haired goat with simple, re-curved horns was probably introduced by the Celts and both this and the old shaggy-haired goats were found by the Romans when they advanced into Britain in 55 BC.

The Vikings, it is thought, were the next to import goats into Britain. They took in their long boats to provide fresh milk during lengthy voyages white, long-haired 'Telemark' goats, which became established in Scotland during the early days of their settlement, and are said to be the ancestors of the wild, white goats which still inhabit the western isles of Scotland. Others

13

claim that these goats swam ashore from shipwrecked Armada galleons and yet others that they are of Swiss Saanen origin. Each theory could have an element of truth as all wild goats in Britain are domestic goats gone feral and no-one is certain when the first goats were seen on the islands.

In England goats continued to decrease in value and, in the 10th century, were worth less than sheep. Norman law banned goats from forests. The earliest forest laws gave rights to commoners to graze cattle, horses and pigs in the woods at certain seasons of the year but forbade them to graze sheep and goats, because of the damage they did, and this law is still in force today. On the Berkeley estates during Edward III's reign, goats were allowed on manors bordering the Forest of Dean and Mickewood Chase and over 300 kids a year were eaten at the lord's table. A mediaeval writer stated that 'Young Kyddes flesshe is praysed above all other flesshe – although it be somewhat dry. Old Kydde is not praysed.'

Goat meat varies according to the animal's age and diet. Good pasture-fed kid killed before it is weaned at about 3–4 weeks old has white, delicate flesh resembling chicken; at 3–4 months the meat is like spring lamb but the flesh of mature goats is dry and tough and more like venison. The Welsh used to cure goat hams. Pennant in his *Tour of Wales* in the 19th century noticed that 'The haunches of the goat are frequently salted and dried and supply all the uses of bacon: this by the natives is called *Coch y wden*, or "hung venison".' The other parts were cooked in numerous ways as is testified by the many recipes for goat and kid meat in old cookery books.

In Norman times cows were used for draught and sheep were the main milking animal, while goats were only kept on open or wooded grazing inappropriate for sheep. In the Domesday Survey it is noticeable that in places where sheep flocks were small those of goats were large, but in many places both were kept for cheese-making. The 11th century cheese industry especially at Cheddar, was entirely based upon flocks of milking ewes and herds of she-goats. Within the Cheddar Cheese making area, at Chewton Mendip, Chilcompton, Rodney Stoke, Stratton, Pilton, Shepton Mallet, Croscombe and Blagdon, herds of fifty to sixty she-goats complemented the flocks of sheep in the Domesday Survey. Pure goats' milk cheese was also common in the Middle Ages and continued to be so especially in Wales and Scotland. Crofters' cheese and Ross-shire crowdy butter were made from goats' milk until the 19th century. Today it is mainly the French and Swiss who persevere with pure goats' milk cheese; St Marcelin, Ruffec and Pont L'Evêque are all traditionally goat cheeses.

Goats' cheese was also considered 'very profitable for the sick'. A cure for respiratory ailments was to inhale a rotting cheese and cheese placed on wounds was said to effect a cure. Goat's blood was also thought efficacious. Topsell, an early writer on these matters, claimed that 'the bloud of a Goat hath an unspeakable property, for it scoureth rusty iron better than a file, it also softeneth an Adamant stone'.

Other parts were also considered medicinal. Horns placed under a pillow cured insomnia and, burnt, their ashes drove away serpents. Their liver cured hydrophobia and the gall dropsy, or dysentery. Modern medicine is quiet on the benefits of these cures but does endorse the diatetic value of goats' milk. In composition it is the nearest to human milk making it better for children than any other. William Harrison in the 16th century said that:

> the milk of the goat is next in estimation to a woman for it helpeth the stomach, removeth oppilations and stoppings of the liver and looseth the belly.

It has also been used to foster many animals from camels to foals and lion cubs. Goats' milk contains about 4% butter fat more than cow's milk, and is rich in Vitamin A, B1 and riboflavin. It also comes in smaller globules which are easier to digest. This makes it suitable for invalids and those suffering from TB, ulcers and infantile eczema. It was also used as a cosmetic for, according to a Gaelic saying:

> With violets and the milk of goats anoint thy face freely,
> And every king's son in the world will be after thee my dearie.

Because of its evenly suspended, small, fat globules goats' milk does not separate and the lack of cream means it makes poor butter; due to the goat's omnivorous eating habits the milk can acquire a wildness in flavour not to everyone's liking. Although most people's prejudice comes from drinking the milk of she-goats run with the he-goats when the milk becomes tainted by the male goats' overpowering 'bodily odour'.

This odour is produced mainly by glands situated behind the horns. Both sexes have these scent glands but the scent is only activated in the blood stream by male hormones, particularly during rutting. The billy's smell is further intensified at this time by his habit of bending his head between his forelegs to catch his urinary spray. Captive goats sometimes carry this to excess and the ammonia in the urine eventually makes them blind. In old goats the scent can be so strong it carries half a mile. As Topsell sagely writes,

> There is no creature that smelleth so strongly as doth a male Goat, by reason of his immoderate lust, and in imitation of them the *Latins* call men which have strong breaths (Hircosi) Goatish . . . And therefore *Tiberius Cesar* who was such a filthy and greasie-smelling old man, was called (Hircus vetulus) an Old Goat.

Although to humans the billy's smell is offensive, to the nanny it is extremely sexy. She finds it stimulating, it brings on her heat period, makes her twitch her tail in the air, bleat, behave in a provocative fashion and head straight for any male in the vicinity. No animal is as sexually precocious as the goat. Kids have been mated at three months by billies of under a year old and produced young. Leonard Mascall, writing in the 16th century, comments on the goat's excessive lust cautioning that:

> There is no beast that is more prone and given to lust than is a Goat,

for he joyneth in copulation before all other beasts. Seven days after it has yeaned and kidded, it beginneth and yeeldeth seed . . . although without proof. At seven months old it engendereth to procreation, and for this cause that it beginneth so soon, it endeth at five years, and after that time is reckoned unable to accomplish that work of nature. That which is most strange and horrible among other beasts is ordinary and common among these for . . . the young ones being males cover their Mother, even while they suck their milk.

It may be a short life, but is certainly a sexy one and Bewick stated in the 19th century that one buck can easily cover 150 females.

Another curious circumstance connected with the goat's sex life is the influence of daylight. Under tropical skies, where seasonal changes in vegetation and climate are minimal, goats breed throughout the year but, in northern climes, the dates on which rutting begins coincides with the shortening days. In the north of Scotland it starts about 10th August, and in Cornwall in late September.

But the billy's sexual proclivity has been thwarted by civilization. Selective breeding for short hair and hornlessness inadvertently destroyed his libido. David Mackenzie in his book on *Goat Husbandry* written in 1957 describes how:

The male goat's wealth of horn and hair is the emblem of his sex and a natural by-product of his glandular system; it is also the only outlet for protein and minerals surplus to his maintenance needs and the temporary demands of the breeding season . . . the dietetic needs of the hornless short-tailed male unfit him for leadership of the flock, which is the natural social function of the male goat.

Such problems did not worry the peasants of the 13th century. What concerned them most was the argument that raged between those who wanted to run goats in the forests and those who wanted to keep them out. Under Scottish law if goats were found in a forest three times the forester hung one from a tree by the horns and, if four times, he killed and disembowelled one on site. Robert Bruce, however, ordered that goats were to go unmolested in the area of Pollochthraw near Inversnaid in Perthshire. Here he is said to have been protected from pursuers, in 1306, by some wild goats who lay down in the entrance of the cave where he was hiding giving the impression that no one was there.

It was about this time that the Bagot arrived at Bagot Park in Staffordshire, and became intertwined with the family history. The Bagot family is one of the oldest in England and owned land before the Norman Conquest but exactly when and how it acquired its first goats is uncertain. Long-haired, horned goats with the same distinctive black head and shoulders, white body and hindquarters are called Schwarzhal goats in the Canton Valais of Switzerland and in the Rhône valley and one theory is that some were collected by Sir John Bagot returning from a crusade through

16

the Rhône valley in about 1377 either as a souvenir or as a source of milk for the journey.

The smart black and white Bagot goats must have made quite an impact at the time, as most goats fitted Turbervile's description of wild goats in *The Noble Arte of Venerie or Hunting*, 1576.

> The Wilde goat is as bigge as an Harte, but he is not so long, nor so long legged, but they have as much fleshe as the Harte hath . . . They have a great long beard, and are brownish grey of colour like unto a Wolfe, and very shaggie, having a blacke list all alongst the chyne of their backe, and downe to theyr bellie is fallow, their legges blacke, and their tayle fallowe.

But compared to sheep, pigs or cows, goats counted for little and according to Harrison in the 16th century were very localized:

> Goats we have plenty and of sundry colours in the West part of England and amongst the rocky hills. Some also are cherished elsewhere for the benefit of such as are diseased or have sundry maladies, unto whom, as I hear, their milk, cheese and the bodies of their young kids are judged very profitable.

These herds often numbering as many as a hundred were controlled by a goat-herd whose wages were supplemented by an allowance of milk at certain times of the year (which he made into cheese) and often by a kid in spring. But controlling a body of such independent animals was an almost impossible task in spring and autumn. For most of the year domestic goats are amenable to direction. But in spring, immediately after the kids are born, and in autumn, at rutting, the king billies' blood runs high, their assertive instincts are strong and herds become difficult to control; scattering in every direction, they head for freedom. In a very short time they revert to the wild and become 'feral'. As a result although no truly wild goats exist in Britain there are many feral goats some of which have survived free and undomesticated for centuries. In the 17th century wild goats were to be seen in Tredegar Park in Monmouthshire where 'Squier Morgan' had 'in his parke a thousand head of deere, beside wild goats and other cattle about his grounds.' They were also found in the forest of Kingswood in Gloucestershire.

Domestic goats were run with cows in the belief they prevented contagious abortion, although there is no scientific basis for the theory. Goats were also kept in stables, and still are, to 'preserve' the horses 'from many epidemical diseases', particularly staggers. But mostly, until the late 18th century, goats were the 'poor man's cow' providing the average cottager's milk supply and groups were to be seen on every common in England. They were only kept in large numbers in areas unsuitable for sheep. There were more goats in Cornwall than anywhere else in England but they were universal in Scotland and Wales, where people being poor and vegetation scrubby goats grazed the hill land behind every croft and cottage. Their importance in Scottish rural life is signified by the frequency with which Garbh and Garbhar, often anglicized to 'garve and gour' and meaning goat appears in place names. Ardgour, Aringour, Garve and Garbhalt are just a few examples.

Before the clearances of the 19th century goats were more numerous than sheep in Scotland. The number each croft could carry was laid down as two or three dozen but was frequently exceeded. The crofters kept goats for meat, and every part of them was put to use; their horns made drinking vessels, knife handles, musical instruments and even bows by stetching an ox's tendon between the points. Their fat became tallow for candles, their skins knapsacks and containers and ropes made from their hair were said not to rot in water. The ropes of goat hair intertwined with pigs' bristle, used by the St Kildan islanders to suspend themselves from cliff tops in search of gulls' eggs, were heirlooms passed from one generation to the next. Long goat's hair, preferably white, embellished kilt sporrans and, tightly curled, baked and bleached, it was made into wigs for legal heads. The sale of skins in the south bought in money (57,000 goats' skins and 43,000 kidskins went in 1698), as did the sale of surplus stock.

Many goats went to supply 'whey drinking spas'. In the 18th century these were as fashionable as water spas and it was as common for invalids to travel to the Highlands to drink whey as it was for them to go south to take the waters of Bath and Buxton or go sea bathing. Many goats were imported from Ireland and until 1914 the Irish goatherd, proclaiming his wares and squirting jets of milk to prove their validity, was a sign that spring had arrived in many a village on the route from Wales to Scotland.

These large droves frequently contained hundreds of goats and must have been difficult to control as escapes have formed a large portion of the wild goat population of Scotland.

What with wild goats and crofters' goats the whole of Scotland was teeming with the animals. Landlords, anxious to conserve their woodlands, made attempts to restrict the number of goats kept by tenants, but mostly in vain. Other lairds filled with enthusiasm for the ideas perpetrated in the 18th century for improving agriculture were similarly thwarted in their efforts by the eat-all indigenous goat population. Requiring the land for the more profitable sheep they forced both goats and their keepers to vacate their holdings, by fair means or foul, pushing them to poor coastal land or even further, across the sea to America. The goats that went with the crofters found their lives greatly changed. The coastal crofts had no hill grazing and as all available space was needed for crops there was little room for goats. Those that were retained were tethered and living perpetually on the same ground constricted their natural ability to resist worm infestation with the result that their productivity diminished. It was then discovered that cows were more easily contained, fitted better with crops, produced more milk per man hour of labour and so they gradually supplanted goats. But with the removal of goats, gorse and other plants inedible by sheep and a nuisance to man, proliferated on the mountains and moorlands.

In England, the Enclosure Acts of the 18th century restricted the amount of common land and here too it became common practice to tether goats, leading to a build up of infection and lowered productivity.

By the beginning of the 19th century goats had become so rare that it was possible for Cobbett to write in his *Cottage Economy* in 1822 that

> I wondered how it happened that none of our labourers kept goats; and I really should be glad to see the thing tried.

and that

> I can see no reason against keeping a goat where a cow cannot be kept. Nothing is so hardy; nothing is so little nice as to its food. Goats will pick peelings out of the kennel and eat them. They will eat mouldy bread or biscuit; fusty hay and almost rotten straw; furze-bushes, heath thistle, and, indeed, what will they not eat, when they will make a hearty meal on paper, printed or not printed on, and give milk all the while! They will lie in any dog-hole. They do very well clogged or stumped out. And then, they are very healthy things into the bargain; however closely they may be confined . . . Goats do not ramble from home. They come in regularly in the evening, and if called, they come like dogs . . . A goat, . . . will face a dog, and, if he be not a big and courageous one, beat him off.

Although goats largely disappeared from the land in the 19th century, they were still found at sea, providing fresh milk on long sea voyages. It had long been the habit of cargo-carrying boats and P & O steamers to collect a few milking goats in India and pick up replacements at Suez and Malta. Once in England these foreign looking goats were disposed of, some to zoos and some to the passengers. In this way the Indian Jumna Pari goat arrived and so did the Maltese goat. On Malta it was the custom to allow the goat to act 'the part of the milkman in towns and cities carrying its own commodity from house to house, even up stairs on different floors, and being milked before the door of the customer, who thus gets his supply pure and unadulterated.' (Holmes Pegler, *Book of the Goat*). Malta Fever is conveyed through the milk and is passed during mating between goats. At the turn of the century, Malta Fever cost the British Army and Navy stationed in the Mediterranean approximately 75,000 days a year in sickness and some died of the affliction. The disease cleared in the cool wet climate of England and the Maltese goats which came here grew free of the *coccus*.

Another ship-board goat was a drooping-eared, sleek-coated, rich black and tan oriental animal found in various forms in Persia, Ethiopia and Egypt. These goats were crossed with the similar-looking Nubian goat. The first Nubians were brought over from Paris in 1883 where they had been introduced to supply the daily milk needs of a hippopotamus presented by the King of Abyssinia to Napoleon III. The Nubians produced prodigious amounts of milk, about 300 gallons a year, with a high, 5%, butter fat content and were termed the 'Jersey of milk goats', but they were delicate.

19

To increase their numbers and hardiness they were crossed with similar-looking Oriental goats and English goats. The resulting progeny, called Anglo-Nubians, are one of the most popular milking breeds in Britain today, and the most popular in the USA. In the 1900's Nubians became fashionable and acquired aristocratic patrons. The Baroness Burdett-Coutts, the philanthropist daughter of Sir Francis Burdett of Coutts Bank, did much to promote this goat 'tall in the extreme with its long legs, Roman nose, very close coat and pendulous ears', a squashed pug-dog nose, and short tail. They were trained to draw small carriages popular at seaside resorts and among the gentry. Two females were sent from the Sudan by the Duke of Connaught as a present to Queen Victoria and were placed at Windsor on the Prince Consort's farm.

At Windsor the Nubians joined another breed, the Cashmere goat. Sometime after 1828 two were presented to George IV by Christopher Tower of Weald Park in Essex. He had got his Cashmere goats in Paris in 1828 from stock imported from India in 1819 by two merchants, Baron Gernaux and M. Joubert as part of an experiment to manufacture cashmere shawls in Europe. The Cashmere goat's natural home is in the eastern Himalayas where they are kept by nomadic mountain tribesmen living above 15,000 feet. For warmth at such high altitudes these white, long-haired goats, with their high twisted horns, grow a double-layered coat consisting of a fluffy lining of wool covered by a waterproof shield of hair. The finest wool, softer than any sheep's comes from goats living in the highest mountains. Traditionally the wool was combed out in spring over a period of 8–10 days and primarily woven into soft shawls bordered by beautiful vegetable-coloured, curving patterns which became fashionable in England.

Manufacturers tried getting wool from Kashmir, but frustrated in their efforts by politics and the monopolies of Cashmere merchants, decided to by-pass them and introduce the goat. But the Cashmere goat only grows its soft wool under the exacting, dry cold of the Himalayas; in mild damp Britain the wool is coarse. In 1889 a herd was sent from India as a present to Queen Victoria. Unfortunately, on the voyage, an officer attempted to delouse the goats by soaking them with an anti-bug mixture largely composed of paraffin oil. This, coupled with the hot sun and sea air, denuded the goats who arrived without a hair on their bodies.

A similar experiment was tried with the Angora goat which probably shared the same ancestry as the Cashmere. This goat was found in Tibet and is covered from head to hock in a thick, enveloping curly fleece and has a 'baa' like a lamb which led early writers to suppose it was a cross between a sheep and a goat. In the 16th century large numbers inhabited Angora in Asia Minor where they attracted the notice of the Dutch Ambassador at Constantinople. In 1541 he sent a pair to his 'Imperial Master' Charles V. The fame of their wool was further spread by Dutch and English merchants who settled in Angora in the 18th century. Soon Angora goats were found in South Africa, Australia, New Zealand and the United States, their wool made into periwigs, furniture velvets, rugs and 'ladies light dress goods, known as "brilliantines" or "lustres" '. Experiments were tried to breed Angoras in England but neither the soil nor the wet climate suited the wool of the sheep. The Duke of Wellington who imported some from South Africa in 1881 in his park at Stratfield Saye in Hampshire abandoned the experiment for this reason. Today Angora goats are farmed extensively in New Zealand and Australia for the production of mohair.

The introduction of these strange and exotic animals by their aristocratic patrons did much to revive an interest in goats. The first goat show was held at Newton Abbot on 19th May 1875. The following year a goat show was held at the Crystal Palace and from then on they became regular events. The formation of the British Goat Society in 1879 under the patronage of Baroness Burdett-Coutts, the Dukes of Wellington and Portland and the Earl of Londesborough gave further respectability to goat keeping and the establishment of herd books laid down criteria for recognized goat breeds.

Standards were set for the ideal milking goat. She was to have a long, lean head and a lively expression with a big jaw to masticate large quantities of food, a long, silky-skinned neck, a strong, straight muscular back, deep wide-sprung ribs and a long, gently sloping rump to support a capacious, heavy udder with knobbly milk veins, thick, pronounced, hand-sized teats projecting forwards towards the belly and straight hocks to avoid bruising the udder as she walked. The front legs were to be straight and clean-boned but the general line of the goat was allowed to vary according to local terrain.

None of the popular breeds in the late 19th century were British. They were either Indian, Ethiopian or French until their position was usurped by Swiss breeds. These neat, smooth-haired, hornless goats with fatty tassels hanging from their jaws had lived lives interwoven with the people of the Swiss and French Alps for centuries. The goats throve on the high Alpine pastures and their breeding, maintenance and milking capacity were given meticulous attention by the tidy Swiss until the goats reached a peak of perfection none could equal. Every canton of Switzerland bred slightly different varieties of much the same Alpine goat adapting them to their environment and requirements. These prodigious goats became renowned throughout Europe and eventually some were brought to England where their 'kindly natures, polite manners and excellent milking capacity' made them instantly popular. Three main breeds became established: The Toggenburg, a milk chocolate coloured goat with white markings said to be a cross between Appenzell and Chamoisee goats; the white Saanen from the Saanen Valley in the Canton of Berne; and the black-and-white Alpine goat. All were cross-bred with British goats to produce off-shoots known as the British Toggenburg, a large goat with high milk yields; the British Saanen, heavier and leggier than the Swiss Saanen and preferring a soft life on cultivated land; and the British Alpine, a large goat weighing about ten stone, a heavy milker and tolerant of free range conditions.

In the 1980s, partly due to changing circumstances minds changed about the assets of Swiss goats. In the opinion of many they had become too specialized and over-bred requiring special housing and selected feeding, both now expensive, to produce their full capacity of milk. There was a revival of interest in the old British goat. Searches were made in the few places where feral goats still existed, on the outlying islands of Lundy and Isle of Man and the West Coast Scottish islands, in the Cheviot Hills and those of Caernarvonshire and Merionethshire and any other places where they had not been destroyed by farmers or by the Forestry Commission trying to protect their trees. Blue-grey goats with a black eel stripe down their backs were found in the Cheviot Hills, white and grey ones in Wales, the Highlands and Hebridean islands and black-and-white ones in Ireland. Within these categories enormous variation exists, and experts debate over the fine points, but the overall impression is of sturdy, long-haired, long-bearded independent individuals. Their blood is now being used to infuse new strength into inbred breeds such as the Bagot and a recent import from the Channel Islands, the caramel-coloured Golden Guernsey.

Wild goats are difficult to catch, centuries of life in the hills has made them nervous, they bound over fences like deer and are brave to the point where they will tackle dogs – in India goats have been seen to kill leopards.

There is little place for goats today. Apart from a few people with herds producing milk commercially for yogurt most goat keepers are amateurs. They keep their goats in parks, paddocks and back gardens and have improved the animal out of all recognition. The goat used to be an animal with a short lactation producing 50–70 gallons of milk a year with a low butter-fat content, now its lactation lasts two or more years and it gives 150–400 gallons of milk with a high butter-fat content and has a productive life of up to ten years.

Goats cannot really compete with cows, their yield is less and they are more trouble. Goat keeping rises in popularity when times are hard. It received a boost during the First and Second World Wars and recently the practitioners of self-sufficiency have made it popular again. Most of the goats these people keep are one of the Swiss breeds or an Anglo-Nubian and some have Pygmy goats for fun. The goat can thrive on land too poor, hot or steep for dairy cattle, will give 3–4 pints of excellent, nutritious milk a day, can accommodate itself in all sorts of housing, can be kept tethered or free and is an engaging if canny character.

1. Wild goat in London Zoo. A goat such as this one was the first farm animal. It roamed the mountain ranges of the Middle East and was taken into captivity around 8000 BC. Primitive man chose the goat in preference to other animals because it was able to survive on a rougher, scantier diet in a tougher climate and could provide meat and milk for sustenance, skins for clothing, water containers and boats, sinews for bows and fat for light. Goats were essential to man while he remained primitive and nomadic and went with him when he began migrating out of the Middle East in about 3000 BC. Early man held goats in such high regard he made them celestial symbols. Babylonian astrologers believed the Milky Way was a she-goat and that Polaris was a he-goat. The Egyptians worshipped them as Gods and the Greeks as symbols of fertility. Neolithic man introduced goats to Britain and kept more of them than any other animal. The goat's methodical mastications helped Neolithic man to clear the indigenous forest and make space for other farm animals.

2. *Bagot goats, the descendants of those brought to Britain by a 14th century Crusader.*

3. Crofters in Scotland used to keep more goats than sheep and lived close to their animals as can be seen in this painting, A Goatherd's Cottage by W. Simpson, 1832.
4. Queen Victoria's Cashmere goats in Buckingham Palace Gardens by W. Keyl, 1848.

5. *Chocolate coloured Toggenburg and white Saanen goats being fed in a paddock at Rotherham in 1947. Both these breeds came from the Swiss Alps and are among the most common in Britain today. Goat-keeping rises in popularity when times are hard. It received a boost during the First and Second World Wars and recently the practitioners of self-sufficiency have given it another fillip.*

6. *'November: Gathering the Acorns' by the Limbourg Brothers. A page from the Book of Hours commissioned by the Duc de Berri c. 1412. In the Middle Ages trees were the dominant vegetation in Europe and vast herds of pigs lived on the provender of the forest floor. They were tended by swineherds who, in autumn, used to beat down acorns and beechnuts from the trees to help fatten the pigs. In Britain many pigs were killed in November to provide food for the winter and it became known as the 'blood month'.*

7. *A Gloucester Old Spot sow with her piglets. This breed was also known as the Orchard Pig because, in its original home of Gloucestershire, it was fattened on windfall apples. The Gloucester Old Spot is one of our oldest breeds. It is not known when it first became established but it could have been in the 17th century as people then preferred spotted pigs. Today it is considered a rare breed but its hardiness and self-sufficient habits are bringing it back into fashion as housing and feed increase in cost.*

8. *Large, ungainly, slab-sided, slow maturing pigs which were closely related to wild pigs (8a) were the norm in Britain. In the 18th century, small, fat, quick maturing pigs were imported from China (8b) and crossed with the rangy, indigenous animals. The mingling of blood made an immediate change for the better and the most effective cross was that with the local Berkshire pig. The fat, black pig that emerged became the most popular breed in Britain and was exported to many other countries (8c).*

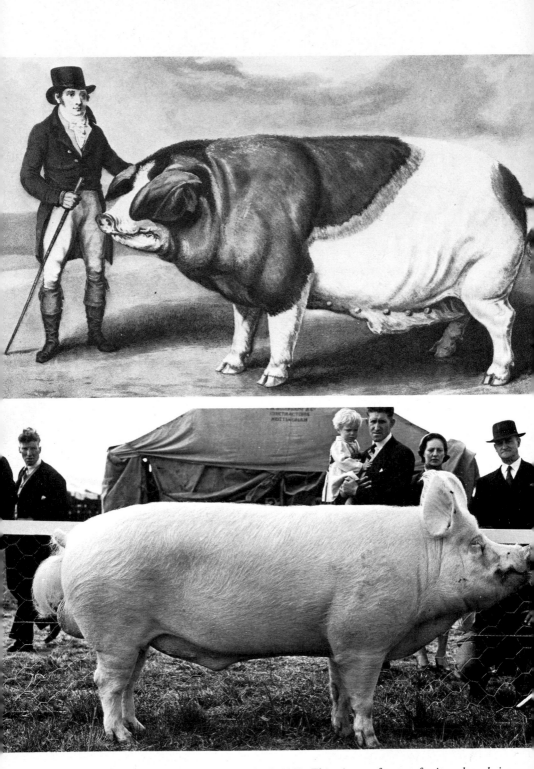

9. The Yorkshire Hog *by R. Pollard, 1809. This pig was famous for its colossal size.*
10. *A Large White boar, 1957. This 19th century Yorkshire breed is now found world wide.*

The Pig

Greed and omnivorous eating habits, coupled with a highly-tuned sense of smell and an adaptable personality, have been the pig's downfall throughout its history. The pig's liking for all kinds of food, from acorns to beechnuts, truffles and roots to any small animal or insect which crosses its path is what led it into human hands. The pig kept to himself while man remained primitive and nomadic but as soon as man began to settle, the pig, tempted by his crops and refuse began pillaging. Unwittingly the pig had found an adversary whose eating habits were as diverse and experimental as his own. And experiment man did, finding in the pig's carcase not only a huge and abundant source of meat, but fat for kitchen, cosmetic and medicinal use, a skin which could be made into leather, and tusks for ornaments. Neolithic man was still coping with the conundrum of domesticating the sheep, goat and cow and it was a little time before he turned his full attention to containing the pig. Remains of domesticated pigs are not found on sites dating from before the Neolithic agricultural revolution in about 7000 BC. From this time on, however, their bones are unearthed wherever there are permanent settlements, throughout Asia, the Middle East and Europe. As domestication advanced the bones proliferate but their size diminishes. This characteristic is common to other animals in the early stages of domestication. The size of an animal is closely related to the quantity of food he eats and early man's primitive agriculture was unable to provide the richness of diet to be found foraging wild.

Sus scrofa, the pig, belongs to the same zoological order as the elephant, rhinoceros, hippopotamus and hedgehog. Early pigs were powerfully built, thickset animals set on high, fast running legs. They were browny black in colour with darker stripes and spots on their bodies and covered in coarse hairs. The hairs became bristles on their thick shoulders and on their heavy, muscular necks a mane which the pig erected when irritated. Their heads were short and broad, their ears pricked and their long snouts were armed with sharp, crooked tusks. Later, two classes of pig began to emerge, a wild forest pig and a domestic, home-loving pig. One was large and fierce, the other smaller and more docile. While forests remained the dominant vegetation the two types continued to exist. The pig, although considered by many to be more intelligent than the dog, is an obstinate animal with a liking for a settled life and a reluctance to be driven anywhere, and with its wide distribution it became easier for man, as he moved from the Middle East into Europe, to stock new farms with local wild pigs. The pig was used not just as food but also for clearing the indigenous forest for farming land. Grubbing their way in search of food pigs uproot seedling trees and gradually destroy undergrowth which is quickly supplanted by grass. It is said that 'pigs were the bull-dozers of antiquity and sheep the mowing machines that followed behind'.

In ancient Egypt the pigs' tasks in agriculture extended beyond bulldozing to trampling down reeds after flooding to make fertile seed-beds and helping tread in the seeds of corn. Wall paintings of the time show pigs being driven into newly planted fields, their pointed hooves treading-in the grain to just the right depth for germination. Once the corn was ripe the pigs were again employed as rudimentary threshing machines to trample the grain from the ears. Some Egyptians ate pork but others regarded the meat as unclean. No one knows at what point, or for what reason, the superstition became established that pig meat is unclean. Some attribute it to the nomads who despised settled farmers and, therefore, their pigs who would not herd or travel. Others opined that their flesh 'although white and delicate, is so flabby and surcharged with fat, as to disagree with the strongest stomachs'. Still others subscribed to what the 12th century Jewish philosopher Maimonides described as the pigs 'extreme filthiness, and their eating so many impurities', 'though he divide the hoof, and be cloven-footed, yet he cheweth not the cud; he is unclean unto you', this led to their meat being forbidden under Levitical Law as food fit for Israelites. It is a fact that the pig is a carrier of the disease causing trichina worm which can be passed to man through undercooked pork. To this day, Jews and Muslims will not touch pork.

The gluttons of the ancient world took a more tolerant attitude to the pig. To the Greeks pigs were sacred, and were sacrificed to the corn goddess Demeter at the beginning of harvest and to Bacchus at the beginning of vintage. The Romans relished pork, made the breeding, rearing and fattening of pigs a subject for study and applied every art to 'impart a finer and more delicate flavour to the flesh: the poor animals were fed, and crammed', with dried figs and honey to bloat their livers 'and tortured to death in various ways, many of them too horrible to be described, in order to gratify the epicurism and gluttony of this people'. The hogs were roasted whole, stuffed with thrushes, larks, warblers, nightingales and oysters and served 'bathed in wine and rich gravies'. (W. Youatt)

The Romans also respected the fighting qualities of the wild boar and the Twentieth Legion made it their emblem. Early cave dwellers in Spain equally revered the boar and as a mark of their admiration painted his image on their cave walls. Some of these paintings, executed in 4000 BC, can be seen at Altamira in Spain. A boar's aggressiveness is based on his sense of social position. Pigs in the wild tend to form natural family groups of ten to thirty sows over-lorded by one boar who establishes firm territorial rights over an area of woodland. Aspiring young boars have to fight their way to the top. Adopting shoulder to shoulder positions and using their tusks as weapons, they shove away at each other for as much as an hour before one submits and rushes into the undergrowth squealing with vanquished pride. Amongst sows there is also a distinct pecking order; any that step out of line are persecuted, often to death. It is this strong sense of social niceties which, when upset by overcrowding in modern intensive systems, leads to perverted and tortuous behaviour amongst pigs.

It was perhaps an appreciation of this natural habit of pigs to form groups of limited numbers that led Celtic law to lay down sometime before 300 BC that a herd should consist of twelve sows and a boar. But the Celts were not the first to keep pigs in Britain. Neolithic man probably brought a domesticated variety of the wild European boar with him when he sailed across the Channel in 2500 BC.

Another theory of how British pigs came to be domesticated is that Neolithic man merely plundered indigenous wild sources. In the densely thicketed forests which covered Britain at this time wild pigs would have been abundant and as Neolithic man became established, created communities and farms which needed tending, and had less time for hunting, the inducement to take pigs into domestication would have been strong. By the Bronze Age in Europe, however, the situation was beginning to change. Bones found among Swiss Lake Dwellings reveal a smaller, longer-legged pig, named by archaeologists 'the turbary pig'. It is thought to be a cross between the wild European boar and a smaller, domestic variety, the result of a haphazard breeding policy which consisted of turning the sows out into the forest to be mated by a wild boar and give birth before bringing mother and young back into the farm-yard. A breeding method which persisted in Britain well into the middle ages.

This 'turbary pig' is assumed to have been brought to Britain by the Celts. The evidence is written in that ancient Welsh poem, the *Mabinogion*.

I have heard there have come to the south some beasts, such as were never known in this island before . . . they are small animals, and their flesh is better than the flesh of oxen . . . and they change their names. Swine are they now called . . . and still they keep that name, half hog, half pig.

Half hog, half pig the turbary remained, as bones dug deep from the archaeological layers of the Celtic Glastonbury village settlements in Somerset show. In other parts of the country in later years there was

25

probably some crossing between the turbary pig and the indigenous wild pig, the resulting off-spring, according to Strabo writing in 30 AD, 'live abroad' in the woods 'and are remarkable for their height, strength and swiftness – indeed it is as dangerous for a stranger to approach them as the wolf'. The Welsh laid down laws making the owner of a pig responsible for any damage done by swine entering a house and scattering the fire. The custom arose of sending pigs into the forest from Midsummer Day on 24th June until January 15th and between these dates the forest was reserved exclusively for pigs.

It was these semi-wild pigs that the Romans found roaming the country when they arrived. It is thought they did little to alter the situation. Unlike other livestock, the pigs' only contribution was meat and, as long as wild sources were abundant, there was no real reason to keep them under proper control. There is a suggestion that the Romans imported some improved pigs of the low-legged, well-fleshed Chinese variety. If so they would probably have been brought by traders along the much travelled Silk Route. The historian Palladius said that six piglets were enough in a litter and that black pigs did best in cold countries and white ones in warm places.

There are two schools of thought on the pig's tolerance to heat. One is that sun reflects off white skins and the other than they suffer badly from sunburn whereas black pigs can stand extreme heat. Both theories may contain some truth. However, pigs going from the grey skies of Britain to the continuous blue of the tropics certainly need a period of acclimatization as they are more susceptible to heat stress than any other farm animal, for they have no protecting coat of wool or fur, merely a thin covering of bristles on a scaly skin. In winter pigs suffer from cold and need masses of bedding and companions to huddle against for warmth and any extreme heat, whether from a build-up of therms in modern-day intensive units or a tropical sun, quickly brings on hyperthermia. Even in a British summer pigs need shade and water in which to make miry wallows and coat their skins with a protective covering of mud. It is said that the beneficial effects of the hot springs of Bath were first discovered when the leprous Baldred, eldest son of the Lord Hudibris, King of Britain, noticed the effect the waters had on the 'scabs and eruptions' of pigs wallowing in the mud.

Sunburn was not something that concerned the Roman pig or its Saxon successor. Shaded by forests, which even after centuries of clearance still dominated most of Britain, they were the corner stone of Saxon agriculture. Pig husbandry formed the principle occupation of the people and vast herds were kept, wealthy men frequently owning thousands of pigs. An earl, Alfred, who died in about 880 and owned land in Sussex, left 2000 swine to his wife and daughter and 400 to other kinsmen. Numbers of pigs grew, but the woodland decreased and competition for the remaining fare became intense. Laws evolved which allowed pigs into the woods only from Midsummer to the New Year after which they moved to grassland. While in the forest they were tended by a swineherd who, during October, knocked down the acorns and beechmast from the trees to help fatten the pigs.

October became known as the month of *porculatio* or fattening swine. In November the fat they had accumulated from their diet of oak and beechmast, crab apples, berries and roots was measured and the pigs divided up into those to be kept as the staple herd and those to be slaughtered. November was known in Saxon times as 'blood month' and the animals were slaughtered when the moon was on the wane as it was thought they ate more by moonlight and would be fattest at the end of that period.

The swineherd slaughtered the pig with a blow from the back of a hatchet. By the 7th century a swineherd was a recognized profession. Each was supplied with a herd of swine, a certain number of which he was required to surrender to the lord every year as payment for woodland 'denbera' and grassland 'gaerswun'. The amount varied from one pig in seven to one in ten and the instigation of the law marked the end of the independent peasant.

This law must have been difficult to enforce. Pigs were universal as place names testify. Barlow meaning 'boar-ground' occurs in Derbyshire, Durham and Yorkshire. Wild Boar's Fell in Cumberland is self-explanatory. In Fife, Muckross is Celtic for 'Boar's Promontory' and the Byro Hills a corruption of Boar Hills. These are near St Andrews, where formally two enormous boar's tusks were chained to the altar in the cathedral commemorating 'an immense brute slain by the inhabitants after it had long ravaged the surrounding country'. The prefix Swin and Hog often applies to an association with swine, while Toller Porcorum in Dorset needs no explanation.

The arrival of the Normans effected enormous changes in the higher echelons of English society but made little difference to the life of the pig. He did spend less time in the woods, evidence that clearances were continuing to have an effect, not entering them until August 29th but still staying to the New Year. The dues the swineherd paid to the lord were now called 'pannage' not 'denbera' but he retained his position of importance in Norman agriculture. The vital part pigs played in early mediaeval economy can be estimated by the fact that they were the yardstick against which all woodland was valued. In the Domesday Book woods are recorded in relation to the number of swine they fed. In Essex out of 465 manors, 413 kept pigs; on the lands of Ely, 104 manors out of 115 kept pigs but in Cornwall, where there was less woodland, there were pigs on only 87 manors out of 243.

Pigs still lived in the woods during the late summer and autumn but, in the New Year, moved to rough pasture and from there progressed to pick the residues of arable crops from those fields and thence to eat the remains of the threshing floor. Sties became a feature in their lives. Early sties were merely enclosures in woods or on feeding grounds where the pigs spent the night, but later sties were built in the courtyard of the manor farms to house farrowing sows and weaklings. Pigs began to be fed on barley, pease and 'drasch' or the spent grain from the home brewhouse and to be fattened for special occasions. A boar for a feast was given ten times more grain and sucking pigs were fed whey and corn by the dairymaid. Sucking pigs were reared on farms where it was not economic to raise the whole litter. They were killed under one month old and roasted whole when, according to Charles Lamb in his 19th century Essays of Elia 'of all the whole *mundus edibilis* I will maintain this to be the most delicate'. He justifies his enjoyment of the pig's premature death by claiming that, 'the innocent' has merely been rescued from the fate of growing up 'to the grossness and indocility which too often accompany maturer swinehood. Ten to one he would have proved a glutton, a sloven, an obstinate, disagreeable animal, wallowing in all filthy conversation – from these sins he is happily snatched away.' Others would disagree with Lamb's character assessment. Gilpin in his *Forest Scenery* of 1791 wrote more sympathetically that

> The hog is commonly supposed to be an obstinate, headstrong beast, and he may perhaps have a degree of positiveness in his temper; but if properly managed he is, or may be, an orderly docile animal. When your meanings are fair, and friendly, and intelligible, he may be led with a straw; nor is he without his social feelings, when he is at liberty to indulge them.

Since the religion of mediaeval man was still imbued with paganism and a love and fascination for the dark, mysterious, symbolic side of life, he decorated his high gothic cathedrals with images from his everyday life then, in a fanciful way, gave them religious significance. So the pig to the mediaeval mind retained its biblical association with evil by its ability to root up 'the tree of life'. It also represented fleshy delights and, in Celtic mythology, was a harbinger of bad luck. Children were warned against the 'cutty black sow' and fishermen considered it unlucky to mention swine at sea, especially when baiting lines. In church, when the story of the Gadarene swine was read, they would whisper 'cauld airn' and feel for the nails of their boots in the belief that the devil entered through the tiny, hair-covered hole in the pig's forefeet. The hole is surrounded by six rings which look as though they have been burnt or branded into the skin and are thought to be the marks of the devil's claws.

Conversely, there exist countless stories of good pigs, of saints who have been guided by pigs to the right foundation sites for their churches and monasteries: St Kentigern founded St Asaph's Cathedral in the Elwy Valley in north Wales where he found a wild boar preparing the ground with his tusks. At Winwick in Lancashire a pig carved in a stone on the tower of St Oswald's church depicts one who, every night, would demolish the day's building work and carry the stones to the spot where St Oswald actually did die grunting 'Winwick, Winwick' on the way.

During the early middle ages boar hunting was considered excellent sport by most of the nobility. There was plenty of game as, in spite of advances in domestication, the forests still teemed with wild boar. The Anglo-Saxons hunted on foot armed only with a boar-spear and aided by powerful 'boarhounds'.

None but the largest and oldest boars were hunted, and these afforded a very exciting and often dangerous sport, lasting for many hours; for when first the animal was '*reared*' he contented himself with slowly going away, just keeping ahead of his pursuers, and apparently caring but little for them, and pausing every half-mile to rest himself, and give battle to his assailants, who are, however, too wary to advance upon him until he becomes very tired; then he takes his final stand, and dogs and hunters close around him, and a mortal combat ensues, in which the beast eventually falls victim. (W. Youatt)

Such numbers of boar were killed that William I introduced preservation laws limiting boar hunting to the month of September. Numbers still dwindled and eventually the wild boar became extinct. The last boar in Essex is said to have been speared to death by Elizabeth I's favourite, the Earl of Essex, and James I killed the last remaining specimen in Windsor Park in 1617. Charles I tried to reintroduce the breed and went to considerable trouble and expense to procure a wild boar and its mate from Germany. They were set loose in the New Forest where they begot large numbers of offspring, locally called 'forest pigs'.

In the 13th century wild pigs, although still abundant, were decreasing in numbers. Walter of Henley, that stalwart mediaeval writer on farming matters, might reserve most of his praise for agile pigs with the ability to root and fend for themselves but, for others, well-bred boars who produced fat, early maturing offspring were more to be recommended. Ideas on selective breeding were beginning to glimmer in the mind of mediaeval man but for the moment pigs remained slab-sided although some were described as having lop ears instead of prick ears. As the woods decreased, theory, and with it captivity and a controlled environment became entrenched in the pigs' way of life. Only in well wooded parts of the country did life continue in its old routine. Each day the village swineherd 'commonly gathereth' the villagers' pigs 'together by his noise and cry and leadeth them forth to feed abroad in the fields'. The swineherd was responsible for any damage the pigs might do if he lost control and as recompense each owner paid him ½d. a quarter for one large pig or two small ones and gave him an annual dinner. The pigs returned each evening to be 'styed uppe' by their owner from sunset to sunrise. Where woodland had largely disappeared, and pigs had to be fed rather than allowed to forage for themselves, method began to creep into the practice of pig-keeping. Records from a farm near Stevenage of 1274 show them to have had a pig unit of a boar and two sows. The sows produced twenty-five piglets a year and the boar piglets were castrated at six months. Although pigs still scavenged for most of the year, at eighteen months they were finished for the table on a supplementary diet of 'pea chaff' – the threshed-out pods of peas. Throughout the country, on the more advanced farms, the science of special feeding for fattening was gaining acceptance. The additional feed came in many forms according to area and personal means. On a Hampshire manor farm and at the Abbey

29

of Bec at Combe in 1306–7 porkers were fed with both 'beremancorn' and 'brotecorn' but at a farm near Wellingborough the pigs only got chaff.

The age at which pigs were considered fat enough for killing also varied widely. Pigs could be slaughtered at any age but generally did not reach bacon weight until they were two years old – the average baconer today is ready at six months. Most pigs met their end in November to provide meat for the winter, although some met a more festive fate as provision for feasts.

Walter of Henley, airily asserted that 'three times a year ought your sows to farrow' a dimly feasible proposition which allowed only six or seven days between giving birth and mating. The author of an anonymous mediaeval Husbandry advised two litters a year, or at the most, five in two years. A figure most modern pig-keepers would agree with. Of the piglets born 15% do not survive weaning; squashing of piglets by the sow's unwieldy weight, chills and disease all take their toll and, out of an average of ten born, only about eight reach adolescence. In mediaeval times disease was a great hazard. Losses due to 'murrain' that all encompassing word used to cover every disaster, except theft and slaughter, were bewailingly common. The other universal disease of pigs was the 'measles', a skin disease probably akin to that affecting humans and thought by the ancients to be a form of leprosy, it was contagious and one reason why pigs were abhorred. The disease was seldom fatal but it made the meat pale and flabby, unabsorbent to salt and generally unwholesome. Some of the cures for measles were bizarre in the extreme: in Lisle's *Husbandry* he recommends isolating the pig for three days and nights without food then feeding it, at intervals, cored apples stuffed with brimstone. Markham advised administering red ochre and urine while Tusser advocated mere isolation and then cavalierly suggests killing the pig for soused bacon for sale to the Flemings, whom he obviously disliked.

Yet more dominant than measles in the everyday life of the pig, as the 15th century ended, the 16th began and England emerged from the Middle Ages and entered the era of Elizabeth I, was the decimation of forest land. In the 13th century there were still over sixty separate forests in England and parts of the midlands were so thick with trees settlers avoided them. Even in the 15th century an estimated four million acres of woodland existed, but by the 16th the destruction of trees which, in those days, were used not only for building houses but to provide fuel and implements, and for fuelling the charcoal furnaces for smelting iron, lead and glass, enraged the inhabitants of Worcester who complained that 'As the woods about here decay, so the glass houses remove and follow the woods with small change'. Camden in the late 16th century noticed that salt works had recently been responsible for the demolition of Feckenham Forest in Worcestershire. No one thought of replanting and no one anticipated the consequences this loss of woodland would have on all including the pig. Indeed the pig itself contributed to the loss of its habitat by guzzling up all the seedling acorn and beechnut. The Masting season was cut to six weeks, lasting only from Michaelmas, September 29th to Martinmas, November

11th, after which the pigs were largely kept in sties 'lodged on the bare planks of an uneasy cote'. They were fed on barley, malt, oats and pease and given the washings of ale barrels or whey from the dairy to drink. Poor cottagers, who relied on their pigs to provide food for the household and could not afford extra food, gathered what they could from the forests and hedgerows – elm leaves, hips and haws, sloes and crab apples.

The cottager treated his pig with scant respect considering how vital it was to his life. Little attention was paid to breeding and the average pig remained a half wild animal, huge and high-backed, standing about four feet at the shoulder, his seven foot long back covered with bristles 'prest down' and was kept in filthy conditions. Given the chance pigs do have a keen sense of cleanliness and will divide even limited quarters into a part for sleeping and a part for manuring. But given too little space this natural behaviour pattern becomes upset, the pig abandons any attempt at keeping clean and does wallow in his own filth. Cottagers in the 16th century too often thought that any place was good enough for a pig and, in fact, many believed that the filthier the pig the better the pork.

Once pigs were contained their rooting ability was no longer an asset. The ringing of pigs at large became general practice and then law. One, William Hull, was fined twenty pence for keeping three unringed pigs 'which subverts the lands of their neighbours to the great damage of their neighbours'. Early rings consisted of a peg of holly passed through the nose, but later copper, iron and white wire rings were used and the task frequently performed by the village blacksmith. He pushed the ring through the portion of bone dividing the snout and nasal bone so that it pressed painfully on the pig's nose if he tried to use his snout to any effect. The gentler method of ringing used today is to insert a wire through a hole made in each nostril, this gives just a tweak if the pig roots.

Ringing could, in some ways, be said to mark a divide in the history of the pig. Never again was he to be a free-ranging wild animal with an independent life of his own. From the 17th century on, the pig's life became increasingly circumscribed, so that today to see a pig loose in a field is a rare sight while to see herds loose in a forest is unthinkable.

Some farm animals can definitely be said to have influenced the development of their environment, but the reverse seems true of the pig. He appears always to have been at the receiving end. As the woods disappeared he was housed in a sty, and as supplies of beechmast and acorn vanished he was fed pulse foods and leftovers from dairy and distillery. Without complaint he swallowed all, converting them into luscious pork and bacon. The pig is, indeed, as William Youatt describes, 'a perfect cosmopolite' adapting itself to almost every climate and food and full of intelligence.

Gilbert White in one of his letters from Selborne mentions a 'very sagacious and artful pig', while Dr Johnson is described by Boswell as saying that 'the pigs are a race unjustly calumniated. *Pig* has, it seems, not been wanting to man, but man to pig. We do not allow time for his education: we kill him at a year old.' Darwin agreed with Johnson. In his *Zoomania* he writes 'I have observed great sagacity in swine, but the short lives we allow them, and their general confinement, prevent their improvement, which would otherwise probably equal that of the dog.'

From the early middle ages and until the 20th century and the development of intensive pig farming, pigs were not generally regarded as commercial animals. In fact in the 17th century pigs could be said to have gone down in social status. In mediaeval days pigs were the preserve of Abbots and the Lord of the Manor, but in the 16th they became the cottagers' animal. Celia Fiennes endorses this drop in social scale by describing in her *Tour of Britain* how in an age when the aristocracy were really privileged and the 'Proud Duke of Somerset' had outriders to clear the roads of peasants, lest they should see him passing, one farmer refused to be stopped from looking over his hedge and held up his pig so that he should 'see him too'.

Pigs were kept chiefly to provide food for the winter and each cottager owned an average of four. Most people had some rights to feed their swine on common pasture and in local woods and fallow ground, but otherwise they fed them as they could according to their means. In town poor people kept pigs in part of their houses or in a sty. Saturday, in Norwich, was the day allocated for cleaning the sties and the pigs were allowed to run about the streets from noon to evening which must have made for some muddles. The only places where pigs continued to be kept on a large scale were those with considerable forest keep, dairy by-products, or regions growing a high proportion of pulse vegetables. It was in these areas that the first elements of any coherent breeding policy were formulated. One of the first to attempt specialized breeding was John Evelyn, that ultimately civilized 17th century man, who 'sent a Portugal boar and sow' to Wooton in Surrey. 'They greatly increased; but they digged the earth so up . . . that the country would not endure it . . . but they made incomparable bacon'. These parts, with the special feeding they could provide and the resulting excellence of the bacon, gained a reputation for their pigs. Fuller writing in the middle of the 17th century says that,

> Hampshire hogs are allowed by all for the best bacon, being our English Westphalian, and which, well ordered have deceived the most judicious palates. Here the swine feed in the forest on plenty of acorns . . . which, going out lean, return home fat. (*The History of the Worthies of England* – ed. 1840)

In this description lies the foundation of the Hampshire breed of pig. Though Hampshire hogs of this time varied little in looks from their forebears, still prick eared, but with rounder, less bristly bodies, thicker thighs and shorter legs, they were on their way to becoming the black pig with a white belt, taken by immigrants to America (1825–35), where they became extremely popular throughout the country. Their lack of back fat led them to be termed 'Thin Rind' pigs. The breed died out in Britain and when, in the 1960's new supplies of this docile, prolific, hardy pig were required they were brought in from the States.

A pig's colour in the 17th century was not so much something of aesthetic and fashionable value as a matter of prejudice. White pigs were preferred, pink pigs were distrusted as 'most apt to take the measles', black pigs were unpopular – perhaps for the same reasons as today, that in some breeds the meat dresses blue not pink – but black spotted pigs were condoned. Could this thought have been in the minds of the first breeders of the Gloucester Old Spot. This pig was also known as the Orchard Pig because, in its original home of Gloucestershire, it was bred to give two litters a year and then to fatten rapidly during the short orchard harvest on windfall apples. The Gloucester Old Spot also drank heavily of the washings from the dairy in this great milk-producing area, the whey – the liquid drained from cheese curds – and the skim milk left after the cream is taken for butter, often thickened with barley, oats or pea-meal. These formed the mainstay of its diet, to such an extent that it was believed they would 'fall off' if given any other food. The modern Gloucester Old Spot closely resembles the old native type, living in peaceful obscurity in the remote lower Severn valley it remained out of the mainstream of breeding and was never contaminated by any new blood.

The other great pig producing areas of the 17th century were the midland counties of Leicestershire and Northamptonshire. In Leicester today there are still special pork butchers and their pork pies are the best in the land. Formerly every farmhouse, hunting lodge and inn in the Shires had its own pork pie recipe consisting of different combinations of chopped pork, herbs, and spices suspended in a jelly made from boiling the pig trimmings with a little apple and encasing it all in elaborately decorated pastry. Melton Mowbray pies were especially famous, but all the pork pies owed their excellence to the fact that the animals in this area of clayland were fed on quantities of beans and pulses. A farmer called Loder, who left records of his fattening techniques, used 20 bushels of beans to fatten five hogs in 1612 and 6½ bushels of beans to fatten 10 hogs in 1613. His best hog weighed 140 lb when dressed.

Loder expended all this food and thought on pigs which were, in modern terms, scarcely worth the effort; large, ungainly, slab-sided, sandy-coloured animals in whose pulse-fed forms the beginnings of the 19th century Tamworth breed can perhaps be discerned. The Tamworth, like the Gloucester Old Spot, escaped adulteration with Chinese and foreign blood. Take away the modern improvements which have rounded up its body, and look at the long snout – long enough in some cases 'to pick a pea out of a pint pot' – prick ears and quick personality and you can easily imagine a rangy, old-English pig. Stories of the Tamworth's origin are legion but it seems most probable that their red colour derives either from a boar sent from India early in the 19th century as a present to Sir Francis Lawley of Tamworth in Staffordshire or one brought in by Sir Robert Peel, MP for Tamworth at the beginning of the 19th century, from his estate in Barbados. The Tamworth is still famous for its red hair and tough characteristics but recently, due to falling numbers, blood lines became narrow and inbreeding led to a reduction in prolificacy. New blood was imported in the shape of three Australian boars in the 1970s and birth levels are back to 17th century figures.

The Tamworth head is a fine one and, when it was traditional to have a boar's head as the centrepiece of the Christmas table, it was often the one chosen. The decorated boarshead at a feast has a long history. A 'wyld bore's head gylt within a fayr platter' was the first course of the feast at Queen Margaret's wedding to James IV of Scotland in 1503. It also formed the annual rent paid by some Benedictine monks to the hangman. The fee was paid on St Vincent's Day on January 22nd, when the boarshead was carried by a solitary monk to the hangman. The head would have had a long preparation; boiled with herbs, glazed and finally decorated with trimmings of lemon and orange peel and prunes stuck in the eye sockets.

Most of the pigs from these midland counties were sold in London to provide meat for sailors at sea. Passed on, no doubt at one of the growing number of markets to which they were slowly driven at the rate of between six and ten miles a day down 'hog ways' by drovers who needed exceptional skill and patience to move the protesting herds. Celia Fiennes describes arriving in Norwich down a 'very steep descent' which led into a 'space for a market to sell hoggs in'. This exchange of pigs further contributed to the opening of people's eyes to the possibilities of breeding, special feeding, and eventually to the importation of the Chinese pig.

The Chinese pig, or rather the various varieties of oriental pig which began to make their way into Europe in the late 17th century, were to create a watershed in the evolution of pig breeding. In China pigs had been the most important domestic animals since Neolithic times. The pig's ability to live in a confined space and produce large quantities of meat were treasured in that closely cultivated country. In China the pig was never despised, model pigs were laid in the hands of the dead and buried with them as it was thought the departed should not be without this valuable source of wealth in the next world. Tradescent Lay describes how in China the swine are cossetted and rarely driven or made to walk but taken from place to place in a kind of cradle suspended on a pole and carried by two men. The difficulty of getting the pig into this conveyance is overcome by placing the cradle

> in front of the pig, and the owner then vigorously pulling at 'porky's tail', and in the spirit of opposition the animal darts into the place they have prepared for him. At the journey's end the bearers dislodge him by spitting in his face. (W. Youatt)

The Chinese pig did, over the centuries, become small – standing no higher than a horse's hock – and rotund with prick ears, a dish face, downy coat and with a propensity to fatten early in life. The popular dates for the arrival of the Chinese pig in Britain are between 1770 and 1780. Whether their introduction coincided with a time when farmers were receptive to new breeding methods or whether the Chinese pigs were brought in to meet a demand is speculative but the mingling of the compact new blood with that of the large, rangy English pig began to make an immediate change for the better. With interbreeding, practically all pigs became the result of some cross between the Old English and the new Chinese pig or one of its closely related cousins – Siamese and Indo-Chinese pigs called Tonkey pigs after their district of origin, Tonquin. From this time on pigs were systematically bred to be smaller and fatter.

Of all the crosses made between Chinese and English pigs the most effective was with the local Berkshire pig. The original Berkshire was a tawny red or white pig spotted with black; it had a thick close body, short legs and lop ears and, although called the Berkshire, was generally found throughout the midlands. By the end of the 18th century the Berkshire crossed with the Chinese had spread throughout most of the rest of Britain. It has even found its way into literature. The tyrant Napoleon in George Orwell's *Animal Farm* was a 'large, rather fierce-looking Berkshire boar . . . not much of a talker, but with a reputation for getting his own way'. But most Berkshires have quick, engaging personalities more akin to Pigling Bland's love, Pigwig, described by Beatrix Potter as 'being a perfectly lovely little black Berkshire pig with twinkly little screwed-up eyes, a double chin and a short turned-up nose' always cheerfully singing and chatting. Today the Berkshires have neat black bodies, dish faces, prick ears, white socks and white tips to their tails; they are early maturing and do well on all types of food and terrain, adaptable qualities that have led to the Berkshire becoming established in Australia, New Zealand and the United States. In Britain selective breeding made it a specialist pork pig and it suffered badly with the rise of the bacon market in recent years. The Berkshire became inbred and numbers decreased. Recently, however, its ability to grow large and put on meat not fat, which means it can be killed at any weight, or whenever market prices are high, have led to a revival in popularity. Some boars imported from Australia broadened the bloodlines and numbers are rising.

In the 18th century there were still no other recognizable breeds as such, but only 'improved' types of pig. Sir Walter Gilbey in *Farm Stock of Old, 1910* writes that 'There was little difference between some of these varieties, which were named after the counties with which they were more or less identified.' Because of the low status of the pig few records were kept, contemporary illustrations are vague and generalized, and all that can really be deduced is that pigs were gradually changing in shape and colour.

The evolution of the pig has always been closely allied to change in public demand and in the 18th century this demand varied from region to region and was heavily influenced by the available feed. All around London an intensive pigmeat industry developed based on the leftovers of the distillery companies, which kept large herds of up to 600 pigs to fatten on the spent grains. In 1740, it was estimated, that 80,000 pigs were fattened in this way, and farmers of the Home Counties petitioned Parliament against what they considered unfair competition. The pigs gained flesh rapidly on this diet although, according to William Youatt, 'the meat is not firm, and never makes good bacon'. He also counselled brandy distillers to dilute their refuse with water otherwise the animals 'become giddy, and be unable to keep their feet'.

In other parts of the country the dairy and its washings continued to be a force in pig keeping, particularly in the great cheese-making areas of Wiltshire and Gloucestershire. A ratio of one pig to each milking cow was kept. In Northamptonshire and Leicestershire pigs were still fattened to a considerable size on pulse vegetables and they made prize bacon. These pigs had lop ears, some so long they trailed along the ground, and were the ones chosen by Robert Bakewell, the great 18th century breeder, for his experiments in pig improvement. He succeeded in making them a lighter-boned, small-eared, quicker fattening pig. Woodland, where it existed, provided a source of food especially for the cottagers' pigs, which were now always ringed and even fitted with wooden yokes to prevent them breaking through hedges.

Where there was no common ground or woodland the cottagers kept their pigs in a shed close to the house. At the distilleries the pigs were kept 'clean and sweet', and at Woburn some model farm buildings were designed on almost 20th century principles. At Wenham Coke at Longford, Derbyshire, 'a stream runs throught the sties, and the meat is given through the wall, without going in among them, from a cistern at one end of the outward yard'. In Lincoln, according to Daniel Defoe, because of the prevalence of monastic ruins the sties were built in ecclesiastical style with stone walls, and arched windows and doors. The norm was somewhere in between these styles and generally consisted of a yard with a trough of peas and a hovel for shelter.

The inhabitants were as diverse in form as their shelter. Every county in Britain contained a different kind, there were white pigs, black, brown and spotted pigs, long-legged and short-legged pigs, ones with prick ears and ones with lop ears and others with super lop ears which 'almost trail along the ground to make way for their noses'. There were Nazeby hogs, Cumberland hogs, a Lincolnshire hog with a curly coat and a gaunt, ungainly 'greyhound' hog from Ireland with wattles hanging from its neck. In Montgomery the pigs were described as looking like 'alligators mounted on stilts' and in Cornwall as 'wolf-shaped'. The Orkney and Shetland island pigs were as tiny as terrier dogs and called 'grice mites'. They ran semi-wild over the islands, and their bristles were made into ropes from which the islanders suspended themselves over the cliff edges in search of gulls' eggs. From this diversity of pig shapes, as Dorothy Hartley so aptly tells in *Food in England*, comes our

> present diversity of ham and bacon . . . the varied characters of the
> orchard pig, the moorland pig, the wheat-land pig, and the forager pig.
> These pigs had character before they were pork. Thus we get Wiltshire
> bacon, the York ham, the Devon and Somerset bacons, the Suffolk

flitch, the Norfolk, and Lancashire, and Durham bacons. All these pigs had some definite reason for their diversity.

The amount of food required by a pig to reach bacon weight varies considerably in the different breeds and strains. For every 1 lb of live-weight gain, a German Large White requires 2·9 lb of starch and takes 108 days to grow from 66 to 220 lbs, while a Berkshire cross requires 3·4 lb of starch and 158 days to reach the same weight. Accordingly, today, pigs are divided into the quick-maturing pork and bacon strains, and the slow-maturing heavy-hog strains. Pork and bacon pigs, given a high protein diet, gain weight rapidly up to 100–150 lbs after which they tend to start putting on fat; heavy hogs lay down weight slowly on cheaper, lower protein food, do not run to fat and are killed at about 200 lb. Modern pig-keepers choose their breed according to their system of farming and the feed available. In the early years of the 19th century, however, everybody liked their pigs fat. This factor, coupled with the newly learned techniques of controlled breeding, led to some colossal pigs. The most famous was The Yorkshire Hog 'a stupendous creature' who at four years of age weighed 1,344 lbs, measured 12½ hands high and almost 10 feet long. His attendant, Joseph Hudon, announced that 'He would feed to a greater weight were he not raised up so often to exhibit his stature.' Pigs that were allowed to run their full course of fatness often needed props in order to stand.

Cobbett advised that 'About Christmas, if the weather be coldish is a good time to kill. If the weather be very mild you may wait a little longer, for the hog cannot be too fat.' Other 19th century writers recommended a frosty morning in November or when the wind was blowing from the north. Whatever date was chosen, the unfortunate animals was starved for at least twelve hours beforehand, then unceremoniously upended on a stool and, amidst protesting screams, his throat was slit by the local pig-sticker. 'To kill a hog nicely is so much a profession that it is better to pay a shilling for having it done, than to stab and hack and tear the carcass about', says Cobbett – and kinder to the pig.

Once dead, the pig was immediately stripped of its hair, either by being enveloped in barley straw which, set alight, singed away the bristles or by scalding with boiling water to loosen the bristles which were then scraped off. Next the pig was disembowelled, the trotters and head removed and the body butchered into sections. Meat filled the house and the family feasted on it for months. The parts were eaten in strict rotation; first the innards made into faggots, sausages, chitterlings and black puddings, then the pork joints – the 'griskins', blade-bones, thigh-bones, spare-ribs, chines, belly-pieces and cheeks – and after them the hams. These had lain for weeks in salt and spices before being smoked and hung for storage from the kitchen ceiling. A ham was always eaten at Christmas dinner and the boar's head used as the centre-piece for the groaning table. From Christmas until spring the family ate boiled pork, bacon and ham supplemented with pies made from the remains. Gaps were filled with such delicacies as ragout of pigs' ears, pigs' pettitoes – made from the fried heart and liver – urchins, a form of haggis and brawn from the boiled trimmings. Cold brawn makes a shining mould, and slices were served for lunch with baked potatoes, pickles and mustard sauce. The fat from the pig was made into lard used for cooking and cosmetics, the skin went to make bags and shoes, and the bristles, brushes. In short every part had a function; the pig truly deserved its reputation as 'the cottager's friend' and certainly was 'the gintleman that pays the rent'.

However, times were changing. As the 19th century progressed and the Industrial Revolution caught hold, increasingly the cottager became the small street dweller. He carried his old ways with him when he moved to town, continued to keep his continuous food supply in a house at the bottom of the garden. But eventually offensive smells and the danger of disease led to the passing of a law forbidding pigs to be kept near the house. This legislation applied to town and country-dweller alike, forcing both to change the habits of centuries. The removal of the traditional right to keep a pig is said to have brought about the greatest dietetic change in the history of England. It also had other repercussions, the cottager was now forced to shop for the food he had formerly grown himself and this marked the beginning of a shift from self-reliance to dependence on others.

Laws are made to cope with the extremes of society often to the detriment of the norm. Certainly many cottagers kept their pigs clean and sweet and became very fond of the friend at the bottom of the garden. Ralph Whitlock tells a story in *The Land First* published in 1954 of 'A dear old cottager of my acquaintance' who 'once showed me the accounts of a fat pig he had recently sent to market. They gave a credit balance of three shillings.
"Not much profit there, Stephen," I ventured.
"No, there idn, is there?" he agreed. "But there! I had his company for six months!" '

This sentiment would be echoed by others who have appreciated pigs for more than their meat. In the old days pigs frequently provided sport and amusement. A pig with 'greasy ears and tail' used to be chased 'by the men and bwoys drough White Horse Vale' as part of the annual festivities connected with the White Horse of Uffington. Their running abilities were thought by some to be much superior to horses' as 'a pig race takes up the yale afternoon, and t'Leger don't take four minutes'. Their traction ability was also put to use by some. Daniel in his *Rural Sports* records the antics of an eccentric farmer who, one day in October 1811, drove into town in a carriage drawn by 'four large hogs'. They had been in training only for six months and Daniel remarks that 'it is truly surprising to what a high state of tractability he has brought the pigs'. The farmer entered town

at a brisk trot, amidst the acclamations of hundreds, who were soon drawn together to witness this uncommon spectacle. After making the tour of the Market Place three or four times, he came into the Wool-Pack Hotel yard, had his swinish cattle regularly unharnessed and taken into a stable together, where they were regaled with a trough full of beans and swill. They remained about two hours, whilst he dispatched his business as usual at the market, when they were put to and driven home again, multitudes cheering him.

Pigs' tractability is endorsed by Pennant. During his *Tour of Scotland* in 1771–5 he states that 'in that part of Murreyshire which lies between the Elgin and the Spey it was by no means unusual to see a cow, a sow, and two young horses yoked together in a plough, and the sow was the best drawer of the four'.

In the past man also used the pig's highly developed sense of smell to his own advantage. Mr Edward Toomer, gamekeeper to Sir Henry Mildmay in the 19th century, trained a pig, along with his pointer dogs, to find and point game in the New Forest. Pigs are famous as truffle hunters. The master turns the pig loose in a wood where truffles grow and watches and waits. When the pig looks like grubbing up a truffle, the master runs up, drives away the pig and digs up the truffle.

Such nice pursuits belong to an age with time and money. As the years of the 19th century passed most farmers tried to follow some course of systematic breeding, although in some cases it was rather a zig-zag course. The natural prolificacy of the pig with its ability to produce a new litter every ten months means that within five years, six generations can be born. Youatt worked out a series of calculations which proved that in ten years ten sows and their offspring can produce, between them, 39,062,500 pigs. From 1770 to 1900 the numerous permutations that were tried on the cross between varieties of English pig, Chinese and Italian Neapolitan led to the hereditary make-up of the average British pig being more complex than that of any other domestic animal. As a result,

no British breed of pig can in its present form claim any great length of ancestry. To a far greater extent than in any other type of farm livestock, virtue lies in strains within a breed rather than with any one breed itself. It is for these reasons, the extraordinary heterogeneity of blood and the variation between performances of families within a breed, that the modern pig progeny testing schemes are likely to be a far more valid guide to merit than breed points; and it is these reasons which make the pig judging ring so farcical a scene on the stage of the agricultural show. (Trow-Smith, *A History of British Livestock Husbandry*)

Throughout the 19th century breeds were made at the whim of experimental farmers. The Black Dorset, Coleshill, Cleveland, Crinkhill, Molland, Middlesex and Devon all came and went during these years. Their main fault was 'an ample development in front, but with hindquarters deficient'. The breeds that survived fitted the tastes of the new consumers who preferred lean meat to fat and would, according to a journalist of the time, James Macdonald 'in the shortest time and at the least cost produce the maximum amount of lean meat in the best parts, with the minimum of low-priced meat and valueless offal'. The ideal pig was therefore 'long, deep, square in the hindquarters, light in front, clean in the jowl'. Out of the maze of miscellany some pigs equating to this ideal shape emerged, the individual characteristics fixed by careful breeding and defined by the new Pig Breed Society which came into being in 1884. They now recognize seventeen different breeds of British pig. In spite of all the crossing, pigs were still closely allied to areas. Interbreeding might occur but local characteristics survived and in many cases dominated. Yorkshire and the east coast produced excellent breeds of white-coloured, pork pigs; the Small White, now extinct, its derivative the Large White and their intermediary, the Middle White.

The Large White was the result of breeding by Joseph Tuley, a Keighley weaver and pig fancier. He produced a pig whose large lean bacon carcase caught the attention of other farmers at the Royal Windsor Show in 1851. Since then the Large White has never looked back as a breed and provider of fine hams. These were originally smoked with sawdust from York Minster and founded the reputation of York ham. 'No breed' to quote one of *His Majesty's Livestock Returns* 'has done more to raise the world standard of pigs.' Since 1851 the pigs have travelled far from York and Tuley's backyard – where they were scratched for half an hour each dinner time and 'washed after t'maister with Saturday night's soapsuds' – and have gained acceptance in the USA, Australia and, most recently, China.

The Middle White, now enjoying a revival of popularity as an early maturing pig for the pork trade, betrays a strong Chinese ancestry by its snub nose and liking for comfort. It grows quickly, producing a higher proportion of flesh to bone at an earlier age than other pork breeds.

In Essex the consolidating ground for modern breeds based on an old 'sheeted' or white belted pig is found. The old Essex pig was famous for its resilience to the reedy wastes of that area. An Essex squire, Mr Weston of Felix Hall, crossed an Italian Neapolitan pig with local pigs and founded the Large and Small Black breeds. They have lop ears which obscure their vision making them easy to control and a black skin which suits them for oriental climes. The disadvantage of their black skin is that it blues the meat, a facet disliked by fastidious modern palates suspicious of anything out of the ordinary which has led to both Large and Small Black becoming rare. However, a combination of the Essex and another southern neighbour, the Wessex, the British Saddleback is now very popular. The Wessex is a saddleback, a black pig with a white belt, which evolved in the late 19th century, a dual-purpose pig, similar in type to the Essex, it is found throughout the southern region of England.

Further west there were more white pigs, the British Lop in Cornwall and the Welsh in Wales, while across the midlands the black Berkshire and red Tamworth still held sway. Both these breeds and all the others based on the Old English, slow-maturing, pig suffered a decline as the quick-maturing bacon pig became popular.

Until well after the First World War the life of the pig continued much as before. But as human populations grew, and land became more scarce, the omnivorous and adaptable pig capable of existing on brought-in food was pushed increasingly into houses to make way for cattle and sheep which are more dependent on grass. The highest proportion of pigs were still, however, kept in traditional pig-producing areas; those with a high proportion of root and vegetable cash crops where the surplus can be fed to the pigs. The old dairy districts also kept their quota of pigs. Cheshire cheese is seldom now made in Cheshire farm houses but the pigs, that were once kept to eat the whey, remain. For the same reasons pigs are most numerous in south Lancashire, West Yorkshire, East Anglia and around London where, in some cases, they are still fed on distillers' waste.

Pigs are now big business. The days when it was said that 'the average pig keeper was too pig-headed to become pig-minded' are over and the homegrown cottage pig is a rarity. Cottage pigs still exist in some places and had a revival during the Second World War. The homegrown pig is also still kept on small holdings and allotments, especially those in Cornwall and Lancashire, where in 1960 on 1,400 small holdings under 1 acre were found 24,000 pigs.

Also in a minority are farms keeping pigs as a sideline. By law they now have to send their pigs to the local abattoir to be killed, the ritual slaughter and preservation of the parts at home is no longer feasible and hams hanging from the kitchen rafters are seldom seen. But on these farms you will sometimes still find regional breeds. For instance, Berkshires are still bred in Berkshire and a few are kept by the Archbishop of Canterbury, Robert Runcie, following, he says, in the tradition of other pig-owning ecclesiastics. 'Thomas à Becket was a great pig-keeper, but I hope I don't share the same fate.' (*Daily Telegraph Magazine*, Feb. 3, 1980 'Road to Lambeth'). The Welsh too is still found in the hands of its traditional owners, the miners of the south Wales valleys.

Many old breeds are also kept alive by enthusiasts interested in them and their history and by scientists unwilling to lose gene source material. The same scientists who, in many cases, have plotted their own and the pigs' downfall, destroying the animals with the very systems they devised. The controlled environment house, the intensive unit with its carefully modulated heat, light and food, where every detail of the pig's life from birth to death, is programmed, often by computer, came into being in the 1950s. At an experimental farm in Suffolk

Friday is farrowing day, every week. The vet calls each Thursday at 8 am and gives an inducing injection to all sows which are known to be within five days of farrowing naturally.

It all happens in the following 24 hours. Or nearly all. Most of the births take place on Friday morning.

The pigman is on duty in the farrowing shed to give any help needed, dock tails, cut teeth and spray navels.

The advantages of timed farrowing are invaluable, says Mr Turnbull – You have a man on the spot for every birth. You can swap pigs around between litters of 13 and 14 and others of 7 and 8. You can size your litters. You can eliminate bullying at the teat. Every pig has an ear number, so you can record accurately.

Because all the pigs are born on one day we can wean all on the same day 14 days later. (*Farmers Weekly* – Sept. 21, 1979)

The piglets are then pushed into wire mesh cages in three tiers surrounded by plastic sheeting, the air carefully filtered to reduce the 'risk of airborne infection' and 'ensure ideal conditions for the piglets'. At 14 weeks they move into fattening pens and at 185 to 190 days weighing 71 kg deadweight from a diet of barley, soya meal, animal protein, whey and vitamins, they are killed.

Their mothers have longer lives but no more comfortable. Attached to the floor by a belt and chain secured round waist or neck they live their lives in rows between metal partitions staring straight forward, able only to stand or lie or eat, their sole function that of producers of further piglets for the battery cages. They are served one week after their piglets are weaned by the cold method of artificial insemination and even separated from any real contact with their piglets by metal farrowing rails, just in case one should be squashed in the confined space and reduce the pig producer's profit margin by a tiny percentage.

The threat of disease to animals whose protected environment has allowed no build-up of resistance always hovers on the horizon. Because of this, artificial insemination and embryo transplant are the only permitted sources of new blood. Original and replacement stock are sometimes even obtained by hysterectomy. The sow is slaughtered just before giving birth and the piglets removed from her uterus which is regarded as

a near sterile container, in which the piglets are free from pneumonia and related rhinitis organisms, swine dysentery, Aujeszky's disease and a range of external parasites including mange. (*Farmers Weekly*, Pig Production Supplement, 11 July, 1980)

And what of the fattening pig? Bored to distraction by his monotonous diet of pellets, in his twilight, centrally-heated, airless, spaceless world, his natural behaviour frustrated, he develops sinister habits – tail biting, ear-biting, bullying of companions and in extreme cases cannibalism. To prevent this, pigs now have their teeth and tail cut at birth; they are sedated on drugs put in their food and induced to sleep for twenty-three out of twenty-four hours by being kept in darkened houses.

All the joy of life has gone for the pig. Artificial insemination even prevents the boar indulging in his unique method of intromission. His penis has a spiral tip which, as the boar thrusts his way into the sow's vagina, literally 'screws' itself in. Only when the spiral glands of the penis are tightly lodged in the sow's form cervix is intromission complete and only then is the boar's sperm released. It flows in high and low waves for from three to twenty-five minutes while the boar rests his head and forefeet on the sow's back. Stockmen call this process 'treading' from the imprint it leaves on the sow's back and look for the marks to see whether she has been properly mated.

Sows, or gilts, as female pigs are called until two years old, first come into heat at six months old but generally receive their first mating at between seven and nine months. A heat period lasts from one to three days and is recognized by a swelling of the vulva and the sow's noisy behaviour. Boars are usually eight months old before being used for service and are given no more than two sows a day or eight a week. A sow releases about twenty eggs during ovulation, most of which will be fertilized. A high proportion, however, die or wither away before the sow gives birth sixteen weeks later. Under natural conditions, before labour a sow will build herself a dry den or nest lined with whatever is available, chewed undergrowth, grass or straw. Ten to fourteen piglets are normally born, of which the sow will rear only as many as she has teats. Piglets adopt a teat immediately after birth and will never suck from another. One or two piglets are usually lost from diseases or are squashed by the sow and an average of eight or nine survive to be weaned at about eight weeks. Normally the sow is served again during her first heat, which occurs some three to eight days after weaning, or at her second a month later.

Given a life reasonably adapted to their natural, gregarious, investigative, curious, greedy, guzzling, rooting, intelligent, self-reliant, hierarchical selves, sows continue to produce about three litters every two years for up to twenty years. Frustrated in their natural habits they 'wear out' in four years; their legs weakened by constant standing on concrete, their digestive systems ruined through the lack of variety and roughage in their diet.

Modern farmers are discovering to their surprise that they cannot regard the pig merely as 'a perishable item' to be manipulated at their convenience and scientists are discovering, through 'research', what neolithic man knew from observation and commonsense, that pigs prefer and do better living under reasonably natural conditions. It is amazing that the idea put forward by Dr Colin Whitmore, head of Animal Production, Advisory and Development Work at the Edinburgh School of Agriculture in the statement

> I believe firmly that provided we can *afford* to improve the pig's welfare at various stages of its life cycle, we will also improve its productivity, and that this will repay investigation on a broad front

could ever be an issue, but that is what modern progress in pig terms is all about.

Such progress is for the future, meanwhile there are mouths to be filled and profit margins to be maintained and if the meat is pale, muscleless and watery, who can remember when it tasted otherwise. An abhorrence of fat, and the housewives' reluctance to spend money on food, brought on the selective breeding which has left both native and imported breeds 'longer and stringier'. In the 18th century pigs were bred so fat they needed support to stand, in the 20th they are too long and lean for their legs. Since the greatest profit comes from pigs with the longest, leanest backs they are bred for this characteristic. Mixed marriages occur, each parent carefully selected for a choice attribute. Thus a Large White will be crossed with a Landrace to produce supremely prolific offspring. These will be put to a meaty boar such as a Hampshire or Belgian Pietrain. The numerous, hybrid progeny of this union are guaranteed by their breeding to grow speedily into the porker pigs supermarkets love to stock and the public have little option but to buy. The overbreeding that can occur from these complex crosses which lead to poor meat quality and 'stress death' are only just beginning to be considered important.

Even so, today a commercial pig will most likely be a hybrid, his immediate ancestry made up of the four most popular modern breeds; the old Yorkshire Large White, now longer and leaner in conformation; the Danish Landrace, a quick-growing bacon breed of great length developed in Denmark from native pigs in the early years of this century with such success that Danish bacon and Danish Landrace pigs are one of the country's main exports; the Welsh pig, another 'Landrace' or 'pig of the land'; and the Hampshire. Such hybrid progeny, their length extended by having one or two extra ribs bred in to their backs, provide plenty of lean bacon rashers; their backs are covered by the thinnest layer of fat and their weight is concentrated on the valuable back hams rather than the poorer quality forequarters. Their breeding is big business and their progeny and meat are exported all over the world. In 1970 there were just over 7 million pigs in Great Britain, 14% of all livestock.

Short of everyone becoming vegetarian, the popularity of the pig, with its ever-ready source of meat able to be created in close conditions, is ensured. An upturn in the pigs' standard of living is not so certain, but just as tower blocks have gone out of fashion for humans, perhaps for the same social reasons intensive units for pigs will disappear. As long as enough of the older, stabler breeds and strains survive, and the pig's natural productivity continues, the hysterical results of science can always be reversed and that jolly character emerge again who, in return for scraps from our tables has, over the centuries, served such succulent fare.

The Sheep

Sheep have changed. The nice but pliable ninny of the average field was once a bright, independent animal living wild in the mountains of Western Asia. The multiple breeds that now exist all began with one, the Urial. A sheep with the form of a deer, the size of a Shetland pony and with thick, strongly wrinkled horns which curled round its cheek, like enormous earphones. In about 9000 BC the Urial was hunted by man throughout the mountains that range from Iran to Afghanistan for its meat and probably for its skins as protection against the intense winter cold. By 6000 BC it was beginning to be herded on the plains of Central Asia and it was becoming domesticated; a fact shown by some bones found in the Belt Caves of Northern Iran and the remains of some woollen cloth at Catal Huyuk in Turkey. One thousand years later some archaeological remains at Uruk in Mesopotamia indicate that there were already different kinds of sheep. Some, it is supposed, were variations on the Urial while others included characteristics found in two other wild breeds, the Argali and the Moufflon.

The Argali's natural home lay to the east of the Urials in the mountains that extend beyond Afghanistan through Tibet and North China. It was similar looking except for its very short tail and it is the ancestor of many Indian and Chinese breeds. The Moufflon lived to the west in the mountains which fringe the Fertile Crescent of Mesopotamia and run into Turkey. It was a small, light, fawn-like sheep with delicate upright horns and a short tail and is the ancestor of many European breeds.

Over the next thousand years the Mesopotamians mixed and refined these three sheep and with domestication the sheep changed in looks. Their tails lengthened, their ears ceased to be pricked and became lopped and they developed a Roman nose, another feature of domestication and probably connected with a shortening of the jawbone as their methods of feeding changed. Their horns grew smaller and completed no more than a single curve due to a weakening of the horn base. They also changed colour; wild sheep generally have short coats made up of a mixture of wool and hair and have brown bodies with white underparts and white, brown or black head and leg markings but at the first step of domestication, for some unexplained reason, they became a uniform brown.

It was these new-looking brown sheep that were taken by the people who wandered out of the Fertile Crescent in search of new lands. Some went east to India, eventually reaching China, taking with them an Argali-based sheep, while others made their way south-west into Egypt with a Moufflon type. Archaeologists place the arrival of sheep in Egypt at about 5000 BC and their introduction perhaps stemmed from a desire by the increasingly civilized Egyptians for fine wool. Jacob is said to have been the first to practise the selective breeding of sheep for coloured wool. The woolly undercoat of the new brown sheep was whitish, and occasionally mutations occurred of sheep with white patches or spots which Jacob was allowed to keep as recompense, when Laban put him to work as his shepherd for seven years to earn Rachel's hand in marriage.

According to the Book of Genesis, to effect these colour changes Jacob followed a method believed in by primitive people, that the colour of the object the sheep looks at when it is mated determines the colour of the lamb. To encourage white sheep, white cloths were hung in the sheep folds, especially round their whitestone water troughs, as sheep generally couple near their water supply. To produce spotted sheep Jacob took rods of green poplar, almond and plane, peeled away the bark to make 'white strakes' and set them by the water troughs. The 'ring-straked, speckled and spotted lambs that resulted' were placed along with the rods in front of the strongest sheep. After seven years Jacob ended up with not only a larger but a stronger flock than Laban and gave his name to spotted sheep for all time. Jacob then realized the advantage of white wool over brown with its ability to take coloured dyes. By the reign of David sheep with fleeces as white as snow were praised everywhere and the Song of Solomon rejoices in the fact that a mistress has teeth as white as 'a flock of sheep just come up from the washing'. Sheep were central to life at that time; their meat gave food, their wool and skins clothing and material for tents, their dung fertilizer and fuel, their guts sewing thread and their horns needles, mugs and the trumpets blown in war. It is hardly surprising that God came to be called the Good Shepherd and Christ the Lamb of God and that the Bible is littered with symbols and references to sheep.

Sheep were also sacrificed. When Aaron destroyed a ram and Moses 'took the fat, and the rump' and 'burnt them on the altar' they were probably using a fat-tailed sheep. This curious breed, whose buttocks and tail were hugely expanded by special feeding was considered a delicacy. Extreme examples had platypus-shaped tails so heavy they made up a third of the sheep's weight and, to prevent damage, were supported on wheeled boards attached to the base of the tail.

Other Middle Eastern sheep developed incredibly fine wool, evidenced by Jason's discovery of the Golden Fleece at Colchis on the Black Sea. The extreme cold of Western Asia encouraged the people to breed sheep with long, soft wool earlier than in other places. They also began spinning and weaving wool, both crafts probably following from their natural observation that matted lumps of wool often fall from sheep in spring, for felting in all likelihood preceded spinning and weaving.

Fat tails and long wool are both attainments of a settled existence and its extension, the emergence of culture. Rolling stones, on the contrary, gather no moss and during the 3,000 years which passed between Neolithic man's departure from Mesopotamia and his arrival in Britain, he hardly progressed beyond being as much a hunter as a farmer. His reason for moving was as frequently to keep pace with the forest, which contained his game and wood supplies and kept receding as the ice age melted and his goats ravaged the vegetation, as to find new pasture for his cattle, sheep and pigs. Neolithic man's sheep remained as primitive as himself; small, brown, deer-like animals with long legs, narrow bodies and short tails, they nibbled at his heels as he pressed forward through the forests of Turkey, the Balkan peninsula and on through Europe. Here Neolithic man's routes divided, some went south to Spain and Italy, others north into Scandinavia and yet others to the north coast of France facing Britain.

Neolithic man was brought to a standstill at the English Channel but, threatened by warrior tribes pressing north on his heels, he looked for new land and eventually set sail in round coracle-like boats across the channel towards the white cliffs of Dover which would have been clearly visible on a fine day. The Channel was narrower then, as the two land masses were still in the geological process of separating, and Neolithic man probably swam his animals behind his boats. On arrival he travelled through the dense forest that then covered most of Britain in search of clear ground, finding it first on the downs round Avebury in Wiltshire.

Working outwards from Avebury he crossed Cranborne Chase and moved onto the Dorset Downs round Maiden Castle leaving remains to be dug by archaeologists centuries later. The experts are unsure whether some of the early bones belong to sheep or goats as they were more similar in looks. The remains of small kraals on the downs in Hampshire and Wiltshire are thought to mark the beginnings of an embryo sheep husbandry. In those days sheep were valued more for their milk and meat than their wool, but gradually the balance changed. Some plain woven cloth of Soay type wool unearthed from a Bronze Age Barrow in a bog in Antrim in northern Ireland shows that weaving had begun in Britain.

In 700 BC the Celts invaded Britain from Europe and on their farms sheep were valued at the same price as goats, fourpence a head. Every household had a 'sheep cote' with a 'legal' flock of one ram and thirty sheep tended by shepherds who counted the sheep by 'rhyming score'. This custom, handed down the centuries, still survives in far western and northern parts of England. The words have become a meaningless jingle but they are based

47

on ancient Welsh numbers and vary in each part of the country. Most common in the north were: Yan, Teyan, Tethera, Lethera, Dic, Sezar, Laizar, Catra, Horna, Tic; Yan-a-tic, Teyan-a-tic, Tether-a-tic, Methar-a-tic, Bub; Yan-a-bub, Teyan-a-bub, Methar-a-bub, Gigget. When he reached twenty, the shepherd put a small stick in his pocket as a marker, and started again. His sheep seem to have been of three types. One was the original brown Neolithic Soay and another the similar-looking, but occasionally hornless Turbary sheep. This sheep is thought to have arrived with the Celts as remains of identical animals are found in the Swiss Lake Village settlements. The third was the Urial type, originally found in Persia, and its remains are found only in the highlands and islands of Scotland, particularly at Jarlshof in Shetland. It probably came from Scandinavia, brought to Scotland by one of the series of men that crossed the North Sea during the period. These sheep have long, thick horns and are in all likelihood the precursors of the multi-horned breeds.

In the south the little brown sheep were still grazed on the open uplands, driven at night into entrenched enclosures since wolves were plentiful. The Romans discovered these primitive sheep when they over-ran England in the 1st century AD. With their advanced methods of sheep husbandry it is small wonder that once they became established they decided to extend their civilizing fingers, beyond the improvement of road systems, housing and government, to the sheep; practising habits of selective breeding perfected on the sheep of Apulia and Calabria in Italy and the Tarantine breed in southern Spain. In Gaul Roman flockmasters so improved their sheep and fleeces that France overtook Spain and eventually became the most important centre of sheep pasturing and weaving in Europe in the early middle ages.

Although there is no direct evidence, it is thought that the Romans quickly supplanted the scrawny, brown, short-wooled Soay type sheep with the ones described by the historian Palladius as being tall and broad with big bones, long tails and long, soft, white wool, which are thought to be the basis of all the British longwool breeds, particularly the Cotswold, Lincoln and Leicester. These areas were most strongly colonized by the Romans and here they made their biggest settlements, built villas and established weaving and fulling factories and so founded the English woollen industry. The first wool mill is said to have been at Winchester.

In the forest clearings of the land beyond, the major problem was of survival. Large, long-woolled sheep might by the whiteness of their fleece communicate a 'superior cheerfulness . . . to rural scenery' but they were less hardy and less adept at evading the ravages of wolves and other prey. There was some mingling between white and brown sheep, as seen in the Portland breed. Named after Portland Bill in Dorset, the Portland's ancient ancestry is revealed by its tan face and legs, heavily spiralled horns and fine wool mixed with hairs. Today it is very rare.

The Neolithic Soay is also rare today. Its decline dates from the Roman invasion, although it was centuries before the breed was supplanted in the hill areas of Wales and Scotland; the growth of the wool industry and the development of new breeds in the south meant it was pushed further and further from civilization until its only home was the remoter Scottish islands where it was tended by people as primitive as itself. Strangely, in the technological 20th century the Neolithic Soay is enjoying a popularity it has not experienced for nineteen centuries. Its lean meat and its ability to survive on scrub land have suddenly become assets in an age where people worry about cholesterol and need to utilize every inch of land profitably.

Other so-called 'primitive' breeds have come under investigation for the same reason. All these are multi-horned breeds thought to be of Urial descent. But although the Urial has impressive horns it only has two, not four or six. Where the multi-horned factor comes from is unknown but it is now built into the make-up of these breeds. They all have coloured fleeces and hock-length tails, characteristics also found in sheep of this type in New Mexico, Uruguay, North Africa and the Middle East. The first of these breeds probably arrived in Celtic times but increasing numbers came with the Vikings from the 4th century on. It is thought they decided that British sheep were infinitely inferior to their own and that they therefore brought in their long boats the sheep we now call Hebridean, Shetland, North Ronaldsay and Manx Loghtan.

The Manx Loghtan is now indigenous to the Isle of Man; a centre used by the Vikings for raids on the Lake District. The Manx Loghtan isolated on the Isle of Man since the end of the Viking era stayed purebred. Confined to poor hilltop grazing on this windy island it has never grown large and is one of the smallest sheep in the British Isles with a 'goaty' face, long slender legs and shoulders, like those of mountain goats, lower than its hindquarters.

In the 19th century Acts were passed that enclosed the hill tops and encouraged forestry, leading farmers into keeping larger, more profitable breeds on the remaining lowland pasture and by 1976 the numbers had dwindled to 150. The breed was rescued from extinction by the Manx Museum and the National Trust who run a flock on the island. Others have been established and the breed is now recovering strength.

The commonest multi-horned sheep in Britain was the Hebridean. It was prevalent on most of the Scottish islands and the mainland and, because of its striking looks, latterly graced the parks of English gentlemen. Similar in looks to the Manx but less pure, the breed is now composed of a mixture of blood including Soay and Blackface. The true Hebridean has a soft black fleece which greys as the sheep ages, a short tail, pointed ears and yellow eyes.

Danish settlers are believed to have introduced the Shetland as there are similar sheep in Jutland. Although the Shetland is not multi-horned the fact that its original home was Scandinavia is shown in the names attached to the wool. Brown sheep are called 'moorit' or moor-red and grey 'shaela', those with light-coloured bodies and darker underparts 'katmoget', while white sheep with brownish spots on their faces are termed 'milk-faced'.

Fairisle knitting patterns, also said to have come with the Vikings, are traditionally made from the different natural coloured wools; named 'cottony wool' and noted for its fineness and softness the most delicate is said to come from sheep kept on the wettest moor pasture.

These primitive breeds were plucked, rather than sheared, because of the nature of the sheep's fleece. All primitive breeds, like wild sheep, have a winter coat composed of a mixture of wool and hair; in late spring, when the weather gets warmer, the wool falls away leaving a summer dress of hair only, so plucking merely assists a natural process. The selective breeding of sheep with a fleece composed solely of wool that grows throughout the year is not thought to have begun before the invention of shears and the first of these date from the Middle East in about 1000 BC.

The long-woolled sheep introduced by the Romans to the south of England had a continuously growing fleece and it was here, where the climate and terrain were easier, the people receptive to new ideas from Europe and tamed by Roman organization, that the agricultural methods introduced by the Scandinavians had most influence. A system of sheep management interwoven with a complicated social structure began to establish itself during Saxon times. At the top stood the thane, or lord of the manor, and below him various classes of tenants, such as the freeman, who were 'foldworthy' or able to graze their sheep on their own land at night and so have the benefit of their manure, geburs, who were only sometimes allowed to do this and cotsetle, whose sheep were permanently grazed on the lord's land. The village sheep were generally kept in a communal flock tended by one shepherd. Part of his duties was to guard the flock against wolves. These were still prevalent and a menace. King Edgar, in the 10th century, in an effort to exterminate wolves commuted a portion of his tribute in return for a certain number of wolves' heads. In northern parts wolves persisted well into the 16th century. During the middle ages they were bad enough for the shepherd in the *Colloquium* manuscript to describe his duties like this:

> First thing in the morning I drive my sheep to pasture and stand over them in heat and cold lest the wolves devour them, and I lead them back to their sheds and milk them twice a day and move their folds besides, and I make cheese and butter and am faithful to my lord.

Sheep's milk was as important as their meat and wool. Although the milking of sheep is probably as old as domestication itself, it is first documented in the 9th century. Sheep were milked for at least three months after lambing, beginning generally on the 1st May and at first ewes were milked three times a day. Milking always stopped before September 8th, the Nativity of the Blessed Virgin Mary, to give the sheep time to build up strength before taking the ram, and winter's scarce fare. Sheep's milk is said to combine the digestibility of goats' milk with the nourishment of cows' and to be superior to both. The annual yield varied from 7 to 12 gallons depending on the grass. This is low compared to a modern milk sheep which can give up to

150 gallons. Most went to make cheese and butter and the lambs were raised on the residue, the skim and whey. The practice of milking ewes died away as wool gained in value. Lambs deprived of their full share of milk do not develop the same quantity or quality of fleece.

When the Romans departed wool manufacture was suspended but, as the country became settled under Saxon rule, the industry began to recover. By the 8th century there are indications that enough cloth was being woven in England not only to clothe the indigenous population but to provide a surplus for export. Sheep proliferated. The prevalence of place names embodying some compound of 'sceap', the Saxon word for sheep, attest to their universal presence. There is Skipton in Yorkshire, Shipdham in Norfolk and a Shipton in at least six counties including Shipton-on-Stour and Shipton-under-Wychwood in the Cotswolds. As more forest was cleared the water table lowered, and the surface became drier providing more grazing land. Much of the forest was cleared by men of Christian faith whose later brethren were to transform the sheep and woollen industry of this country. In the 5th and 6th centuries Christians began arriving in Britain founding small monastic communities in isolated places. The life of the 5th century Welsh Saint, St Brioc, describes how the brethren 'gird themselves to work, they cut down trees, root up bushes, tear up brambles and tangled thorns, and soon convert a dense wood into an open clearing'. In 697 Wintred, King of Kent gave to the monastery of Lyminge pasture for 300 sheep at Romney Marsh and early in the 8th century the Abbess of Gloucester acquired Pinswell in the Cotswolds for a sheep walk.

The price of wool continued to rise and with it the sheep population and, by the end of the 11th century, they were beginning to overtake pigs in numbers and when the Normans arrived and carried out the Domesday Survey they found 46,000 breeding sheep on Manor land alone. In the natural sheep areas of Sussex, Oxfordshire and Gloucestershire flocks were numbered in thousands and each serf had as many as 50 sheep. In the 13th century the 198 villagers of South Domerham and its hamlet of Merton owned between them 3,760 sheep while their lord's flock numbered only 570.

In spite of their increased numbers and the increased value of their wool, the management of sheep hardly varied. The old and feeble were sorted between Easter and Whitsun, shorn, then fattened on fresh spring grass and killed for mutton on about St John's Day, June 24th. Mutton usually came from sheep over four years old, and was individually flavoured according to the grass. The old Welsh Border raiding song sums it up succinctly with the words 'O the mountain is the sweeter, but the valley is the fatter; And so we deem it meeter to carry off the latter'. Mountain mutton was spicy from the thyme and herb grass of the hills. It was cooked and eaten with accompaniments like thyme jelly and mint sauce which brought out the special flavour of its lean meat. The coarser midland breeds required fruitier, sharper sauces as adjuncts to their fattier meat. Salt-grazed or marsh mutton was different again and eaten with samphire or the seaweed, laver, to complement the distinctive iodine taste of the meat. But mutton, as Lydgate asserts, is wholesome whether roasted or 'sodyn' – boiled – and the broth is beneficial after illness. Every part of the sheep was eaten. Robert Burns elevated the sheep's head with the following rhyme:

Oh Lord, when hunger pinches sore
Do thou stand us in stead
And send us from thy bounteous store
A tup or wether head.
Amen. (A Burns Grace from the Globe Tavern, Dumfries.)

The first written recipe for haggis dates from 1300:

Take the Roppis (pluck) with the tallow (white fat) and parboyle them and hack them small with pepir and saffron, salt and brede (oat bread) and yolks of eyeroun and swete mylke. Do all togederys into the wombe of the sheepe and sethe hym and serve forth.

Shanks Jelly, Battered Trotters are both mentioned in recipé books along with the more appealing fare of Boiled Leg of Mutton with Caper Sauce, Roast Saddle, Haunch and Leg, Breast and Cutlets.

Every other part of the sheep continued to be put to use, the horns and bones were made into knife handles and different implements, the tallow provided light, the guts became harp strings, the head was boiled into a 'royal ointment', the trotters laid as flooring and the fat or lanoline from its wool and around the udder for all manners of salves and dressings.

In southern areas from Martinmas to Easter, ewes and lambs were housed, fed with hay, some straw and pease. Their care was still largely in the hands of the shepherd who virtually lived with the sheep sleeping in a hut beside the flock. His working day was long; in summer it averaged fifteen hours from 5 a.m. to 7 or 10 p.m. with half an hour off for breakfast and another hour and a half in the middle of the day for dinner and a sleep, but in winter it was shorter as he was only required to work during daylight hours. The shepherd's prime quality was that of patience and he should not be 'over-hasty'. Mediaeval shepherds enlivened the monotony of their days with bagpipe and horn and were often called on to play on festive occasions. But too much socializing was not encouraged by their masters who laid down that they should not absent themselves without leave at 'fairs, markets, wrestling matches, wakes, or in the tavern'. Other music that enlivened an often lonely existence was provided by the bells worn by each sheep. The bells varied in size and thickness to produce different tones and from them the shepherd could tell the position of every sheep. The tinkling of the bells was heard over greater and greater stretches of England. Sheep were becoming the 'sheet-anchor' of farming. They were still small with light fleeces but interbreeding between the little brown, short-woolled Soay and the white, long-woolled Roman sheep had produced white, mostly hornless breeds. Attention was being paid to quality, the prices paid for rams varied from fourpence to sixpence indicating that steps were being taken to improve flocks.

Sheep diseases and their cures are first mentioned in the 13th century. Thousands died from the great 'sheep rot' of 1283, especially on marshland. The rot or, as it came to be known, 'liver-fluke' continued and continues to be a cyclical menace of wet ground. The circle is difficult to break, the fluke's eggs pass with the sheep's droppings into wet ground and grow into larvae which then burrow into small snails. Once they develop into worms they leave their host and cling to wet grass, which is eaten by the sheep; whereupon they burrow into the sheep's liver, laying their eggs and destroying the tissue in the process which kills the sheep but not before it has passed the eggs back into the ground through its droppings. Today drainage and sheep dosing largely contains the fluke but, in the 13th century, the cause was unknown although low-lying pasture was suspected and attempts were made to keep sheep off low ground and to change their grazing every few days.

The other devastating diseases of the middle ages were the 'scab' and the 'murrain', a generic term for all unexplained afflictions. Various 'cure-alls' were applied. An early remedy was an ointment made from quicksilver and lard then, in about 1295, tar mixed with butter or lard became the normal cure. From that time on the tar-box was an indispensable part of every shepherd's equipment and any sheep with white veins under the eyelids, a skin that did not redden when rubbed, wool which pulled away easily from its ribs or which, in November, melted the hoar-frost too rapidly off its back in the mornings due to overheating, was rubbed with grease and tar.

In spite of the devastations of disease the flocks grew in tune with the commercial demand for wool. The growth of the Great English Wool Market was well under way by the beginning of the 13th century and the bulk of wool was supplied by peasant farmers. Many of them were beginning to acquire wealth through the sale of wool. This wealth was to allow them to grow in stature and gradually to establish themselves as a separate Yeoman class on an equal footing with the lord of the manor.

In the scramble for wool money, it was increasingly only those without sheep who remained poor. Villein flocks enlarged until they often outnumbered the lord's demesne sheep. Many lords did have enormous flocks, Henry Lacy, Earl of Lincoln owned 13,400 but the largest single sheep owners of all were the monastic houses. In 1259 the Bishop of Winchester had 29,000 and the Priory of St Swithin near Winchester 20,000 grazing on the Hampshire downs, while between them the abbeys of Peterborough and Crowland kept 16,300 in the fens. The monks changed the face of the landscape, draining and reclaiming marsh and fen and bringing hundreds of square miles of new land into cultivation. The most far-reaching changes of all were contrived by the Cistercian houses who formed solitary settlements at Tintern in the Welsh Hills, at Melrose, and Kelso in the Scottish Borders and at Rievaulx, Jervaulx, Byland and Meaux in North Yorkshire. Here they brought wilderness into cultivation, changed the character of large tracts of country, radically altered sheep farming by introducing ranch-style methods and became in the process 'sheep farmers *par excellence*'. Fountains Abbey farmed one million acres in the Craven district of Yorkshire and, with the other fourteen Cistercian houses in Yorkshire, came to produce an annual wool clip of nearly 200,000 fleeces.

And what of the origins of their sheep? It is thought unlikely that such experienced sheep farmers would found their flocks on what local sheep there were in such an uninhabited area. Improved sheep from southern counties were probably imported and records of sheep movements at the time back up this supposition. Some Exchequer Accounts of 1323, sounding like a mathematical tease, give an indication. On Thursday May 12th, John Le Barber, valet of the chamber of Edward II took charge at Long Sutton of

a combined herd and flock of nineteen cows and one bull, 313 ewes, 192 hoggasters, 272 lambs, and one bell wether, with eight boys and a master shepherd as drovers. He accompanied the stock first to Bolingbroke, where 160 ewes and six rams, one shorn sheep, 118 hoggasters, 127 lambs and a second master shepherd and three more boys joined the immense drove and then moved on with it to Waithe where it was further swollen by the addition of twelve cows and one bull, 127 ewes, 107 hoggasters and ninety-three lambs, with a third master shepherd and two boys.

These sheep were in all likelihood Lincolns, or Lindseys as they were frequently known as, at that time, the finest woolled sheep came from the dead flat Lincolnshire levels round that town. The Lincoln is depicted by Ellis in his *Shepherd's Guide* as being the 'longest-legged and largest-carcased sheep of all others, and although their legs and bellies were for the most part void of wool, yet they carried more wool on them than any sheep whatsoever' and, as David Low says of, 'altogether peculiar' quality 'such as no other country in Europe produced'. Hornless with white faces, they were gaunt animals who matured slowly but since meat was secondary to fleece this did not matter. Farmers selectively bred to produce sheep with heavier fleeces of 8 to 9 inches long, with ringlets of fine, lustre wool until eventually a single ram's fleece could weigh up to 42 lbs. In the process the sheep's frame grew ever more emaciated and when, in the 19th century, the tables turned and meat became important the fortunes of the Lincoln declined except in Australia, South and North America and Europe where it was valued as a crossing sire to transmit its large frame to smaller, quicker-maturing breeds.

The forebears of the Old Lincoln breed were the equally large-boned Romney Marsh sheep. This was the first breed of sheep in England to receive a name and was developed by the ancient and well-organized estates of the priory of Canterbury. Sheep have always been stocked at greater density on the Romney Marshes than elsewhere and, over the centuries, they became one of the most resistant to worm infection and foot rot. They

11 & 12. *The wild Urial and tiny, Neolithic Soay, ancestors of our domestic sheep.*

13. **A Highland Landscape** *by Richard Ansdell, 1881 shows a Scottish drover beginning his journey south. Before the railways, sheep were walked to London.*
14. *(below) The Romney Marsh, our oldest breed.*
15. *Jacob ewes and lambs. Spotted sheep supposedly originate from Jacob's biblical flock.*

16. *Herdwick ram. This Cumbrian breed was resuscitated by Beatrix Potter.*

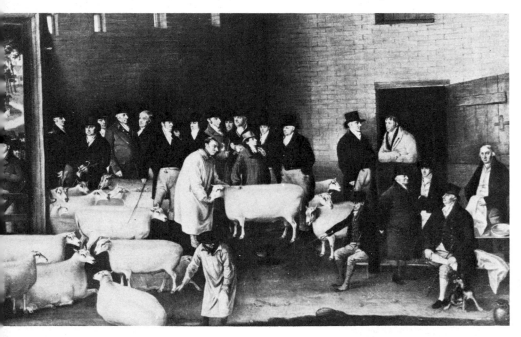

17. *Robert Bakewell hiring out his New Leicester rams for the season, by T. Weaver, 1809.*

18. Shearing sheep by machine. In early times sheep's wool was plucked by hand.

19. Milking ewes in 1981. A modern revival of this once common practice.

20. *Cheviot sheep on their native hills. They were a cause of the Highland Clearances.*

21. The inspection of varieties of cattle in Egypt c. 1400 BC.

22. Chillingham Cattle, B. Bradley, 1895. Examples of wild cattle 'emparked' in 1225.

23. *Oxen harnessed to a cart, survivors of an era when they were the common draught animal. After the Civil War oxen were gradually supplanted by horses.*

also became thorough foragers, feeding evenly and widely over pasture and producing a high cash return per acre. Their blood was temporarily adulterated, by 19th century improvers, with New Leicester, but this softened the sheep and coarsened its fine, close-packed lustre wool. The Kentish farmers, who had 'predicted disappointment' for the improvers, bred back to the sheep which could withstand the 'cutting winds' of winter and the snow-covered marshes destitute of food and shelter.

These 12th century monastic flocks of large, flat-sided, long-woolled sheep formed the foundation of the mediaeval English wool empire. Some of the wool went to France to make fine cloth but most fuelled the mills of Flanders and Florence. On Flemish pastures sheep grew coarse wool but the soil was rich in fullers' earth and suited the cultivation of dye plants. The well-governed Flemings came to rely on imports of food and raw materials which they paid for with manufactured goods. England had no peer as a source of fine-quality wool and, at the peak of their production, the Flemings were prepared to take every ounce. An elaborate system of buying and selling grew up with contracts made in advance for each year's wool clip. The great English monastic houses acted as collecting centres and dealt with the Flemish and Italian buyers.

Before sale the wool was divided into three categories: 'great wool' or sound fleece, 'broken wool' or the sweepings from the shearing shed floor and finally 'shed wool' collected from fields, lambs' wool and the fleeces of scabby sheep. The wool was inspected, to ensure it was 'without cot, gard, black, grey, clacc or villein fleece', that it was untangled, free from loose scraps and droppings and without coloured fibres or diseased fleece. Regulations also covered the conditions under which the sheep was sheared, to ensure the fleeces were clean and free of grease. Many abbeys, including Crowland, had huge thatched wool houses where the clip was sorted and stored, sometimes for years, if prices were low. The wool was wrapped in canvas for the journey to Flanders and Florence. Here the looms turned the fleeces into broadcloth which clothed the entire civilized world. Flanders became so financially dependent on English wool that any stoppage brought unemployment and starvation.

A break occurred in 1297 when, according to the English chronicler Hemingburgh, Flanders became 'well nigh empty because the people cannot have the wools of England'. This shortage was in all probability due to an epidemic of 'scab' which struck with increasing devastation throughout the late 13th century. In 1298 at East Knoyle in Wiltshire 505 out of a flock of 635 sheep died.

A more major break in the wool supply to Flanders occurred in 1336 when Edward III declared war on France. It marked the beginning of the 100 Years War which would never have taken place without wool to provide the finance. Wool, its sale and the fortunes it made induced one of the greatest periods of social and economic change Britain has ever experienced. With it the villein made himself a free man, while honest burgesses climbed upon wool into the ranks of the nobility, only outstripped in their progress there by the dishonest ones, who arrived first like the de la Poles of Hull. The very Lord Chancellor plumped himself down on a wool-sack, and the kingdom might have set on its great seal the motto which a wealthy wool merchant engraved on the windows of his new house:

I praise God and ever shall
It is the sheep hath paid for all.

This social change was accelerated by a disease, not of sheep, but of humans. In 1348 the curse of the Black Death struck for the first deadly time. This bubonic plague which succeeded in successive outbreaks over 30 years in reducing the population by about 1½ million people, or between one third and one half of the total, brought far-reaching reversals in the distribution of both humans and livestock. In many places 'depopulation was such that sufficient labour could not be hired to keep home farm and manor under cultivation even if it was worth it and the landowner was often driven to adopt other expedients. One of these was to abandon any attempt at arable farming and enclose all demesne land for sheep farming' (Seebohm); since fewer men were required to look after sheep than to till arable land. Where plague had not totally decimated a village, landlords often gave a final push by evicting the tenants and demolishing their houses. More than a thousand villages and hamlets vanished in the midlands; their agricultural strips perpetuated as rolling ridges of grass.

As sheep replaced people and grass covered the land, the structural bones of the landscape were shown over mile upon mile of hill and dale the width and breadth of England for the first time in its geological history. The patting hooves and close-cropping teeth of the sheep smoothed the outlines and revealed a new England. Vast numbers were kept in each flock. But as the century progressed 'against a background of pestilence and war, prices fell, agrarian profits dwindled, the population decreased and colonization came to a stop' (Eileen Power) and sheep numbers began to fall. The Wars of the Roses ruined many families, causing the final collapse of demesne farming. The breakup of the largest monastic and baronial estates in the late 14th and 15th centuries made possible the development of the private flock and a new middle class; the Yeoman farmer and the town *nouveau riche*. At the same time the swingeing taxes Edward III imposed on the export of raw wool to raise finance for his continuing war with France, and restrictions imposed by the Company of Wool Staplers who now held a quasi-monopoly of the wool trade, began to affect production seriously. The export of raw wool decreased from 35,505 sacks in 1310–11 to 21,079 sacks in 1447–8 while the export of cloth rose from almost nothing in 1310–11 to 13,425 sacks in 1447. Flemish cloth workers were enticed over to England where they taught their craft to the English who, by the 15th century, had changed from a nation of wool exporters to a nation of cloth exporters. The places where cloth was manufactured were determined by the invention of the fulling mill, a process used to felt finished cloth, giving it a smooth surface. Previously cloth was trampled by human feet but now it was beaten by mechanical means and as the beaters were driven by water power, the new cloth-makers built their mills where steep hills tipped streams fast downhill, congregating on the Pennine slopes, Welsh border hills, Cotswolds, Mendips, Devon valleys and the banks of the East Anglian rivers. Less was heard now of fine Lincoln and Stanford cloth.

The type of cloth produced in each area was decided by the wool the sheep grew in that environment. The character of a fleece is determined by climate, altitude, soil type, quality of pasture and food. The effect of keep upon fleece has always been known, 'their wolles shall grow after the gift of the leys' said a Tudor sage. Cold contracts the pores of the skin allowing fine wool to emerge which matts to defy frost and storms while warmth opens the pores and coarser, less weather-resistant fibres grow. Good pasture increases a sheep's bulk and the length of its wool while poor keep – but not starvation keep – produces lean sheep with short fine wool.

Wool is distinct from hair in that each fibre is lined by imperceptible barbs which interlock when the wool is spun. Each fibre also has a character and strength, softness and springiness – called crimp and closely allied to the curliness of the fibre – shine or lustre depending on the breed and good management of the flock. All these factors are taken into consideration when deciding the method of weaving and final purpose of the cloth. Coarse long wools, which measure up to nine inches, are generally first combed straight then gathered into swatches before being drawn out into lengths as a preparation for spinning and weaving into worsted cloth or hard yarns; while short, soft pieces of wool, measuring about three to four inches, are teased or scribbled into a mass and then drawn out into lengths to be woven in finer cloth. This is then finished by being fulled or felted and its pile raised by teasing (formerly done with the dried heads of the teasel plant) and finally the surface is closely clipped.

Thus Britain, with its incredibly diverse geography, came to have more breeds of sheep than any other country in the world. As cloth-making grew in importance in the 15th century each area developed its own individual sheep and produced distinctive wool cloths. Sheep were not distinguished by breed, but classified by region. The cloth manufacturers on the Pennine slopes in the West Riding of Yorkshire made material from sheep of 'reasonable bigge bone, but of staple rough and hairie'. In the midlands the lustre wools of the Leicester and Lincoln made kersey cloth. The rye-growing area round Leominster in Herefordshire was the home of the incomparable Ryeland sheep whose wool was so valuable it became known as 'Lemster Ore'. From the Welsh border hills came 'friezes', a coarse woollen cloth with a nap on one side only. Wales produced flannel. The best came from the mountainous counties of Montgomery, Merioneth and Denbigh. Here the rich upland pastures made the small wiry Welsh sheep grow the softest wool.

Further south, down the River Stroud, in the Mendips, a 'prodigy of trade' developed, especially in Devizes, from the manufacture of 'drugget' cloth, a woven, felted, coarse fabric used both for cloth and for covering carpets. The cloth was made from the wool of the Wiltshire Horn sheep, an animal valued today for the fact that it does not grow wool. This means it suffers less from fly troubles in hot countries like Australia and has extra energy for growing meat. Selective breeding over the centuries brought about this characteristic and in the 16th century the Wiltshire Horn produced a thick, matted coat. It belonged to a group of massive, white-faced horned sheep from the south-west of England with short to middle length and silky curling wool. They are now divided into three breeds, the Devon Longwool, South Devon and Dartmoor but in the 16th century were collectively known as Bamptons. Their wool was made into mantles.

In East Anglia Celia Fiennes noted that 'the ordinary people both in Suffolk and Norfolk knitt much and spin'. In the lanes round Wymondham you meet 'ordinary people knitting four or five in a company under the hedge'. Other wool went to make worsteds for which the area became famous. These were made from a sheep with a jet-black face; although it is now extinct in its pure form, its descendant, the Suffolk, is one of the most popular breeding sires in the country today.

Kerseys, a coarse cloth emanating from the fulling mills of the Stourside villages of Essex was made all over the southern part of East Anglia. Throughout the region towns bustled with the comings and goings connected with the wool trade.

Such scenes must have been repeated all over England but were most prevalent in the Cotswolds. Here grazed a tall, curly-fleeced Cotswold sheep with a white face crowned by a fringe of wool.

In the 15th century, the Cotswolds became the richest clothmaking centre in the country. The wool grown here, according to Camden, was so 'fine and soft (that it) is held in passing great account amongst all other nations'. The light serge fabric it made created fortunes for its manufacturers who used the profits to build some of the most beautiful buildings in England out of the indigenous limestone. At Northleach, Chipping Campden and Fairford churches, instead of the usual knight resting his feet on a lion, there are effigies of woolmen with one foot firmly on a woolpack and the other on a sheep. Wool and its manufacture entered into the bones of everyone's life, as is testified by the way its terms still permeate our speech. We talk of 'spinning a yarn', 'unravelling a mystery', 'pulling the wool over someone's eyes', 'the web of life' and persist in calling all unmarried women 'spinsters'.

The demand for English cloth seemed never ending. The power and money it generated went to many heads, corrupting some landed barons and, those former leaders of the wool industry, the Cistercians. Henry VIII brought them smartly to heel in 1539 with his Act of Parliament dissolving the monasteries and dispersing their estates. The great retinues maintained by the feudal households and monasteries were left to fend for themselves and their deprivation was further increased by the new middle class of town merchant and yeoman farmers who quickly bought up the old estates. In the entrepreneurial age of Elizabeth I they became as greedy as the landlords they replaced, looked for ways to save labour costs and sought more land over which they would have more control. The enclosure of land was the answer since communal flocks gave their owner no control over breeding or diseases caught from a neighbour's scabby sheep. But as Lord Ernle says in *Farming Past and Present*, it was only after 'considerable tracts of cultivated land were converted to wilderness traversed only by shepherds and dogs where roofless granges and half-ruined churches marked sites of former hamlets that the real social consequences these improvements inflicted on the rural population of parts of the country' were seen. 'Where forty persons had their livings, now one man and his shepherd hath all' was the cry. Although between 1455 and 1637 no more than 750,000 acres were enclosed and only 35,000 thrown out of work (in the 15 years before 1912, 4 million acres were enclosed for pasture), in a population of only 2½ to 3 million it made quite an impact. More, in his *Utopia*, cries that sheep 'that were wont to be so myke and tame, and so small eaters, now, as I heare saie, be become so great devowerers, and so wylde, that they eate up and swallow down the very men themselves'. Their effect on the landscape also began to worry some people. A Nottinghamshire historian in the 17th century bewails the 'numberless numbers of goodley oaks' which had been cleared for sheep 'to graze upon a Carpet Green' while in this century the botanist H. J. Massingham calls sheep 'living lawnmowers' and talks of the 'limited number of grasses and flowers able to exist under their constant nibblings'.

But nothing stopped the march. In the 16th century, once the social consequences were seen, Parliament did pass a series of Acts forbidding the conversion of arable land to pasture, commanding enclosures to be thrown down, demanding decayed houses to be rebuilt, limiting the number of sheep and farms which could be legally held by one man and imposing severe penalties for disobedience to the new provisions. But these were of little avail and hardly any notice was taken by the landlords. A solitary fence driven across newly laid pasture satisfied the statute that it should be restored to tillage. The number of sheep to be owned by one man might be limited to 2000 but the ownership of flocks could be attributed to sons or servants.

The offenders, the sheep, were for the most part large-boned, with long legs to enable them to walk the distances over these extensive sheep runs. They still suffered a great deal of scab and rot, diseases which some writers ascribed to damp summers, rank pastures and 'rowtie fogs'.

In June, before shearing, the sheep were washed by being driven through a stream damned to give depth. A road running through a ford is often all that remains of these 'wash pits where the willow shadows lean'. Two to ten days elapsed between washing and shearing to allow time for the fleece to dry. The shearers John Clare says in his *Shepherd's Calendar* 'wi scraps of songs and laugh and tale' and breaks for 'meats, wafers, cakes and warden pies' took 'the sturting sheep wi trembling fears' and worked until the fleece 'lies beneath the snipping of his harmless sheers'. Later in July the yearling lambs were shorn and from then until autumn the sheep were pastured on grass. In arable areas, once the crop was taken, they were folded over fields, their movements controlled by wattle hurdles moved each evening. This had the effect of feeding the sheep and manuring the land. In autumn the sheep were put to the ram. Fertility then was low and one lamb was the best that could be hoped for from each ewe. Twins were a rarity. Today 1½ lambs per ewe or a lambing percentage of 140% is average and 200% or more considered ideal. A ewe's breeding season, like a goat's, is controlled by the diminishing daylight and they only come on heat when days become shorter than nights, with the result that sheep generally mate and give birth once a year although there are exceptions, such as the Dorset Horn, which will breed and mate at any time.

Unless she is running with a ram a ewe shows less evidence of being on heat than any other farm animal. With a ram she becomes quite forward, seeks out his company, consorts with him for several hours before her heat begins. The ram hardly needs any encouragement. It is a fact that the libidinal forces in species which nature allows only a short mating period are very high during that time. After mating the ewe follows the ram around as long as her heat lasts hoping for more, which she usually gets. Studies show that a ram mounts each ewe about six times during her heat period which lasts anything from one to three days. When the proportion of rams to ewes is low – the average is one ram to thirty sheep – the older wilier ewes are more successful at gaining the ram's attention than the inexperienced maidens.

In the 16th century in some areas the ewes were still being milked after lambing. Tusser says that five ewes gave the milk of one cow but Mascall remarks on the 'poor husbandry in many places where they do use to milk their ewes' where lambs and fleeces were never as good. When the historian Camden rode through Essex in the late 16th century he saw large flocks of milch sheep and noted that:

> there are commonly fed four thousand in this island . . . I have observ'd the young men with their little stools milking them, like women in other places, and making cheese of ewe's milk in their little dairy houses built for that purpose; which they call *Wiches*.

These cheeses, 'great and huge, wondred at for their massiveness and thickness', were sold at market.

Throughout the 16th century sheep were regarded as the most profitable of livestock. However, wool was becoming more difficult to sell. Prices had risen, and Elizabeth's quarrels with France and the Netherlands had closed these important cloth markets. New outlets were sought and eventually found, in Germany and further east in India and the Orient, but they could not compensate for the loss of the major European market and the cloth trade faltered. Acts were passed in 1647 to lower wool prices for home manufacture and further Acts from 1666 onwards forbade the import of foreign cloth. Home demand was stimulated by ordering the dead to be buried in woollen fabric and for everyone to wear woollen clothes on Sundays. Then a new market emerged, for meat.

Towns were growing and changing in character. It became insanitary to keep animals in city centres and no longer were the larger towns near enough to the country to obtain food easily. Supplies had to be brought in and this demand for meat encouraged farmers to change the habits of centuries. These changes were first evident in the counties nearest London. By the 17th century fat sheep were coming off the Romney Marsh of Kent en route for the metropolitan meat market. In 1629 the Toke family, owners of large estates around Ashford in Kent, sent a score of fat wethers to Smithfield indicating the state of things to come. These wethers were castrated rams. Once a ram's masculine urges are removed all its energies go into eating and it fattens faster. Nowadays ram lambs not required for breeding are castrated almost at birth but, in the 17th century a ram could be almost five months old before he was 'cut'. Henry Best, an East Riding yeoman farmer, gave detailed and rather ghoulish instructions for the castrating of rams, describing how a good man could geld a hundred lambs an hour by slitting the scrotum with a long, sharp penknife, drawing out the testicles with his teeth, and dressing the wound with tansy, to deter flies, and butter, to encourage healing. The stones, Best advised, made 'a dainty dish, being fryed with parsley'. This method of castration still survives in remote areas but the more common practice is to fix tough rubber bands round the scrotum to restrict the blood supply, thus causing the testicles to wither and eventually fall away.

Farmers apply a bewildering range of names to sheep according to their sex, age and region or origin. W. Youatt explains some of the variations:

The male is called a *ram or tup*. While he is with the mother he is denominated a *tup* or *ram lamb*, a *heeder*; and in some parts of the west of England a *pur-lamb*. From the time of his weaning and until he is shorn, he has a variety of names, he is called a *hog*, a *hogget*, a *hoggerel*, a *lamb-hog*, or a *tup-hog*, or a *teg*; and if castrated, a *wether hog*. After shearing, when probably he is a year and a half old, he is called a *shearing*, a *shearling*, a *shear-hog*, a *diamond* or *dinmont ram*, or *tup*; and a *shearing wether* etc., when castrated. After the second shearing he is a *two-shear ram*, or *tup* or *wether*; at the expiration of another year he is a *three-shear ram*, etc., the name always taking its date from the time of first shearing . . . The female is a *ewe*, or *gimmer lamb*, until weaned; and then a *gimmer hog* or *ewe hog* or *teg* or *sheeder ewe*. After being shorn she is a *shearing ewe* or *gimmer*, sometimes a *theave*, or *double-toothed ewe or teg*; and afterwards, a *two-shear*, or *three-shear*, or a *four* or *six tooth ewe* or *theave*.

The only way of telling a sheep's age with certainty is by its teeth which only grow in its lower jaw; the upper being composed of a hard, thick-muscled pad. The age is told by the growth of permanent incisors which reach their total of 8 when the sheep is 4 years old. Sheep normally live for 10 years although there are instances of them producing twins at 15. Culley, in his book on *Livestock* tells of a wether aged 20 and this sheep was probably kept alive because it functioned as a 'guide sheep'. Sheep like to follow recognized routes and always use regular watering sites. Another characteristic common to all mountain breeds is that of 'hefting'. This is an instinct which ties a sheep to the place where it was born. A mountain sheep will graze within a few hundred yards of its birthplace all its life. Because of this mountain farms are always sold with their native flock and shepherds always know where a particular sheep is likely to be grazing. Hefting negates the need for fencing but also makes the introduction of foreign sheep difficult.

The movement of sheep is also deeply affected by their habitat. Hill sheep will travel further for food than lowland breeds but the draw of the flock remains constant. Although sheep put onto new pasture initially fan out and move some distance from each other or form sub-groups, they continue to maintain a loose group and move in a regular pattern round their grazing land keeping in touch by 'baaing'.

A sheep mainly grazes at night during the summer and during the day in winter. They eat most intensively at evening and early morning but graze in all for about ten hours in four to seven main periods, each cycle punctuated by rumination, rest and mere idling. Sheep consume 2·5% of their body weight each day and balance their diet, when given the chance, by eating certain plants but apart from disliking hairy or greasy plants they are not fussy feeders and will eat any food. Grass, however, is their main food, eaten by being gripped between the teeth of the lower jaw and the hard musclely upper pad and partly bitten, partly torn from the roots. Sheep bite closer than any other animal.

Their method of grazing also encourages pasture to produce more herbage. The closeness of their bite cuts off short suckers and sprouts making the plant throw out numbers of stronger fresh shoots and the tearing motion loosens the roots of the grass causing them to spread.

This action of making rapid snatches at the grass, clenching it between its teeth, snapping it off before moving swiftly on to the next patch; selecting, snapping, swallowing in quick succession, its head nodding to and fro with the action while its legs move its body slowly along behind had, by the end of the 17th century, deeply influenced the environment of the sheep and this in its turn had influenced the looks of the sheep. There had been advances in the science of agriculture but most of these had happened in the arable world. Livestock remained largely untouched. A tour made of Britain by a man with a discerning eye, as the 17th century merged with the 18th, would have detected minute differences in the sheep. An indication of how differently they developed in different districts can be seen most clearly in the way old breeds are called after areas of country.

A particularly fertile heath sheep, white-faced, long-legged, large-boned and with middle to long wool was found all over the south-west of England

and southern parts of the midlands. In 1732 a Cornish farmer got 19 lambs from 9 Cornish ewes while in Dorset the sheep developed a unique ability to lamb twice a year, in spring and autumn. The sheep population of Dorset became enormous. Leigh writing in 1659 commented that 'within 84 miles compasse round Dorchester three hundred thousand sheep' were to be found. On the Berkshire downs a hornless blackfaced version developed, while in the Vale of Aylesbury there were 'vast numbers of well fleeced sheep' and on the Sussex downs a smaller, stockier sheep. The Romney Marsh still held sway throughout Kent while the old Longwools were to be seen in numbers over the southern midland hills. Over to the east, in East Anglia, and in the Fens, the sheep from the flat salt marshes grew a heavier but coarser fleece and a frame so large that Lincolnshire farmers were advised to use a tup 'of lesser size' on other breeds if they were to avoid fatalities at lambing time. Northamptonshire, Camden remarked, was 'filled, and as it were beset with sheep' and Leicestershire 'feedeth a vast number of sheep' large-boned and of good shape. Sheep with fine wool still inhabited the limestone hills of the Cotswolds and the Ryeland with its superlative wool held its own round Leominster.

A multitude of small heath sheep were found on the Welsh border hills, 'some all black' and many with a great deal of hair in the wool. In Wales, according to Markham, the sheep were 'all of the worst'. In Cheshire there were few sheep 'because their Ground serveth better to other purpose' and according to Smith their fleeces were 'more like brown, such as we call a Sheepe Russet'.

In the Peak District of Derbyshire and in the hilly counties of Lancashire and Yorkshire, picking a sparse living was a heavily horned sheep, some with black noses and with long, straight wool of coarse fibre mixed with hair. In more northerly parts of Scotland, these sheep were supplanted by tiny, horned, short-fleeced, dun-coloured sheep and 'sa wild that they can nocht be tane but by girnis (snares) the (hair) on them is lang and tallie, nothir like the woll of sheip nor gait' (Boece). Sheep remained fixed in a multitude of forms until, at the end of the 17th century, a few enlightened landlords decided to put into practice some of the new theories in circulation. They were lent an incentive by the increasing urban population, the expansion of foreign markets and the establishment of crops which enabled farmers to keep a full complement of stock through the winter. But before any innovations could be fully exploited the over-riding obstacle of communal grazing had to be cleared.

There was little reason to improve land, when one negligent farmer in Gloucestershire, as Lord Ernle wrote in *Farming Past and Present*, by 'not opening his drains, will frequently flood the lands of ten that lie above'. Overstocking could also 'produce a beggarly breed of sheep . . . of little or no value'. Whereas, where enclosures had been effected, 'in fifteen or twenty years, property is trebled; the lands drained; and if the land has not been converted into pasture, the produce of grain very much increased; where converted into pasture, the stock of sheep and cattle wonderfully improved'. Since no individual would attempt to improve his flock while all grazed together, enclosures were the obvious solution. Although they started in the middle ages, the 18th century enclosures were on a far larger scale with farther reaching effects, especially for those at the bottom of the property scale, who were dismissed as 'lazy, thieving sort of people' whose sheep were 'poor, tattered and poisoned with rot'. When land was enclosed, the peasants' individual plots were often too small to support life. Selling to any ready buyer they drifted to town to seek a living in the factories which were just beginning. Deprived of the facilities to grow food they depended on imported meat. At the same time cheap wool began to arrive from abroad, undercutting the English market, and both these factors contributed to swingeing profits to be made at home from wool to meat.

Robert Bakewell was the man of the hour. This Leicestershire farmer saw that the sheep being bred matured too slowly to produce the ready quantities of meat required by the new townsmen. Bakewell capitalized on the pioneer work done by earlier breeders, such as Joseph Allon of Clifton, on the local, long-woolled Leicester sheep. Taking this as his basis he bred in the blood of several types, primarily the Lincoln and Ryeland; one to give size and the other to give a compact form. The characteristics Bakewell sought in his sheep he fixed by a process of in-breeding. After working in secret for between twelve and fifteen years he set the finished article before the eyes of Arthur Young, an ambitious young agricultural reporter. Bakewell's reputation and that of his sheep were made. Marshall's criticisms that the weight of the meat had been pushed to the forequarters where the inferior cuts lay were squashed with the retort that any mutton was of benefit to the poor. Others noticed a loss of wool quality in Bakewell's new sheep but he answered these fears saying that 'it is impossible for sheep to produce mutton and wool in equal ratio: by a strict attention to the one, you must, in a great degree, let the other go'. Mutton the New Leicester did produce and at two years old, whereas other breeds of the day did not until they were three and four years old. At this age the Leicester was beginning to lay on fat 'like the fattest bacon' and only had 'a ready market amongst the manufacturing and laborious part of the community'. The great agricultural experimenter Thomas Coke is reported as selling Dishley Leicester meat 'at 5d the lb. to poor people at Norwich' while feeding himself the lean mutton of the local Norfolk sheep.

The New Leicesters had other disadvantages. They were 'by no means the most prolific, nor the best nurses' and had lambing difficulties. They also needed too much good grazing to flesh out their coarse carcases of lank wool and these eventually led to their downfall. These problems were not immediately apparent. For a while in the 1880s Bakewell's Leicesters reigned supreme and he made a fortune hiring out rams by the season, charging as much as 300 gns for the services of one ram. In his peak year of 1789 Bakewell made 1200 gns from three rams and 3000 gns from his whole letting.

Other farmers followed Bakewell's example; some with more lasting success. Most notable were two brothers from Northumberland, Matthew and George Culley, who took Bakewell's sheep in 1767 and, crossing them with the thin-fleshed, slow-maturing sheep of the Cheviot Hills, produced the white, clean-legged, roman-nosed Border Leicester. It is still one of the most widely used crossing sires in Britain and abroad, making the offspring of mountain breeds such as the Blackface, Cheviot and Welsh Mountain, more fleshy.

Another great improver was John Ellman of Glynde Farm near Lewes in Sussex. Over a period of fifty years he 'directed his attention, in an especial degree, to the improvement of the native sheep of the Downs'. David Low remarks of Ellman that 'He displayed none of the narrow selfishness which, it is to be regretted, appeared in the proceedings of his distinguished contemporary Mr Bakewell'.

> He did not experience the necessity of creating, as it were, a breed, but was contented to adopt the basis which was afforded him in the one already naturalized in the Sussex Downs. He did not carry any of his principles of breeding to an extreme, but acted under the guidance of temperance and judgement.

The new South Downs remained able to pick a living off the scantiest diet and their fame spread over the south of England. From Sussex to the south-west the Southdown was used to improve local breeds. Since the Second World War the demand for larger, heavier carcases has led to a fall in the popularity of this small-jointed animal and it is now classified as a rare breed although it is still popular in Australia and France.

In the 18th century these new sheep met the criteria of the day. The results might not always work, but to men for whom the science of breeding was new, the way sheep could be manipulated to produce profitable rewards was fascinating. Breeding sheep almost became fashionable. George III kept a model farm at Windsor and took an interest in the Merino, importing five rams and thirty-five ewes from Spain in 1792. The Spanish Merino had long been famous for the exquisiteness and denseness of its wool. It grows five times more fibres per square centimetre of skin than any other breed but only under a hot sun. The breed is now found mostly in Australia and the finest wool comes from Tasmania where the climate most resembles that of the Merino's Andalusian home. In cool, damp Britain the Merino wool deteriorated, it became lank and coarse and experiments to breed the sheep pure were soon abandoned, but not before the regal seal of approval given by 'Farmer George' to sheep breeding had stimulated other landlords.

One of the most distinguished and successful was Thomas Coke of Norfolk. When in 1776 at the age of twenty-two, he inherited the family estates at Holkham he set about improving the land and his 'few Norfolk sheep with backs like rabbits'. Farming mainly New Leicester and South-down sheep and anxious to speed up the dissemination of knowledge, which he reckoned travelled at the rate of one mile a year, he started the Holkham sheep-shearings.

These began as a way of gathering together farmers to exchange information and to exhibit his latest rams but grew until, in 1817, open house was kept for a week. Others copied Coke's lead, the Duke of Bedford at Woburn and Lord Egrement at Petworth were two who held similar gatherings. These first 'Agricultural Shows' developed into what are now elaborate yearly events: the Bath and West Show, the Royal Show and the Highland Show. The prizes they offered had an enormous influence on sheep-breeding, giving farmers an incentive to improve their animals. It also made them change their breed. A prize of 20 gns awarded in 1801 by the Bath and West Society:

> To the Stock Farmer who shall have bred and kept . . . the greatest number and most profitable sort of sheep in proportion to the size of his farm, in consequence of his having changed his flock from what had been usually kept . . . in the neighbourhood. (Gilbey – *Farm Stock of Old*)

was won by Mr Duke who had substituted Southdowns for Wiltshires twelve years before and increased his annual profit.

Yet another breed was introduced by large landowners to decorate the parks of the grand houses they were building. Across the 'Ha Ha' which separated garden from park it became fashionable to graze a flock of ornamental sheep. Hebridean sheep with their black coats and soaring horns were considered suitably pretty and a distinct type of this sheep emerged as an exclusive parkland animal. The most popular parkland sheep was the multi-horned, spotted Jacob. Gentlemen travellers doing the 'grand tour' of Europe in the 18th and 19th centuries brought stocks from Portugal and Spain until fifty-nine separate herds existed in parks such as Chatsworth, Somerford Park and Tong Hall.

On their farms, however, these illustrious landlords were busy substituting the often scrawny indigenous sheep with larger, faster-growing Southdowns and New Leicesters. Their lead was followed and soon no self-respecting farmer thought his flocks complete without a dash of New Leicester or Southdown blood. Not always with good results. What happened to the Romney Marsh is a good illustration. The introduction of Leicester blood, although in some ways it refined their rather large, flat-sided sheep not ready for the butcher before it was three years old, at the same time made it too delicate. Then farmers began to breed back to type but not before many valuable genes had been destroyed. Leicester blood also affected the hardiness of some hill sheep.

Mothering power, that intensely strong maternal relationship, that develops between a ewe and her own lamb, was also weakened. So was milking power. A ewe's milk is most influenced by the quality and quantity of her feed and by her age, but some breeds are much 'milkier' than others, and so give their lambs, entirely dependent on milk until four to six weeks of age, a better start in life. Worst of all was the dilution of genes which had taken centuries to evolve and which fitted each sheep perfectly to its particular environment.

In the march of progress such niceties went largely unnoticed. Adaptation to environment no longer appeared vital to breeders who could now grow fodder crops and make sheep to fit their farming patterns. In Dorset meadows bordering rivers were 'drowned' during the winter by a series of dams, sluices and trenches to preserve them from frost, fertilize the grass and enable it to start growing early in spring so that the farmer could produce fat lambs early for the London market. The Dorset farmer was further aided by the fact that he had the only breed in Britain able to lamb out of season and one which invariably threw twins. These handsome-looking sheep have spectacular curling horns and compact bodies covered with a close carpet of wool. Daniel Defoe on his tour in the 18th century was told that 600,000 sheep fed on the light arable land within a six-mile radius of Dorchester. Dorset sheep were sold at the main sheep fairs of the south-west, Yarnbury and Weyhill. Here men gathered from Oxfordshire and Kent to buy the prolific Dorset ewes for breeding fat lambs and the wethers to fatten and sell in London at Smithfield Market.

As the Industrial Revolution progressed, and towns grew even larger, the meat market rose in importance. London was by far the largest consumer of meat. Little rivulets of sheep starting from remote hills in Scotland, forming streams at central 'trysts' or fairs such as Falkirk, swelled into rivers at fairs like Stagshaw Bank in Northumberland and Appletreewick in Yorkshire. From here these sheep moved south in droves down roads cutting straight across country, stopping for the night in recognized fields along the way until they reached Smithfield. Other sheep made similar treks from the west country, East Anglia and Wales.

Drovers owning a dog received an extra sixpence a day. The dogs' duties and looks had changed over the centuries. In the middle ages their main task was to guard the sheep against wolves and other predators, and for this large mastiff-like dogs were used. In the 18th century sheep dogs had to be shaggy to withstand the cold and exposure of their position and have sufficient courage to face a ram. The collie dog, named from the celtic word 'Coillean' meaning little dog, was found in the north. Rough or smooth-coated, usually black and white, very light and swift and of extraordinary intelligence and patience, these skilful dogs are able to round up numbers of scattered sheep on rough ground directed from a distance by the occasional whistle or cry from the shepherd. This art was probably developed in the Scottish borders but the practice took a while to spread.

Although Smithfield and many other markets had existed since the middle ages, the specialized meat market, coupled with the droving system, had a revolutionary effect on farming. In the old days, flocks were static and the sheep, kept for their wool, were only killed for meat when their fleeces or the ewes' ability to breed faded with age. As the meat markets arose sheep became valued as much for mutton as for wool. And this factor, coupled with the new fast-fattening sheep had a dramatic impact on the composition

67

of flocks. A process began of taking sheep in stages downhill. Thus, as a simplification, the offspring of hardy, mountain ewes such as the Blackface were sold at market to farmers in the valleys who crossed them with less hardy but quicker-maturing breeds such as the Border Leicester or, today, the Suffolk. Their progeny, now termed Mules, were either fattened by the valley farmer or, as frequently happens now, sold on for fattening to a farmer with richer grass or arable crops. These 'store' sheep in temporary 'flying flocks' were finally sold to dealers or drovers at fairs near London and eventually reached their demise at Smithfield.

These advances in management had the effect of consolidating the advances in breeding methods. Farmers now began to design sheep to fit them for the market at which they were to be sold. Breed Societies were initiated which set standards for looks, wool and meat quality for these new improved sheep. The sheep of Britain changed. Over the South Downs instead of there being varieties of a kind of short-woolled, heath sheep there were recognized Down breeds created with the blood of the New Southdown sheep. The Hampshire Down was bred by William Humphrey of Cold Ash near Newbury in Berkshire. He infused Southdown blood into local varieties, chiefly the white-faced Wiltshire Horn and speckle-faced Berkshire Knot, and produced the Hampshire Down, a sheep densely covered with wool with the meat-producing ability of the Southdown.

The Dorset Down was also born at this period emerging from a mix of local Dorset sheep, Southdown and Hampshire. It was a sheep with a quality wool, which was used to make socks and fine felts for pressing bank notes. Later, in 1830, the Oxford Down, the largest of the Down breeds, was made from a cross between the Hampshire Down and the Cotswold sheep. The rams are now used mostly as sires on other breeds to produce large, early-maturing, lean lambs able to survive on all types of grass. Leicester blood was used to transform the middle-woolled sheep of Devon into the Devon Longwool, the South Devon and the curly, lustre-woolled Dartmoor. It refined the Wiltshire and Lincoln sheep but had a radical effect on the old Cotswold, altering it from a premier wool breed to one producing coarse, fat meat much liked in the 1850s.

In the west midlands, an admixture of Southdown blood made the Shropshire, Down, Kerry Hill, Clun and Radnor out of a complex mass of the sheep breeds of Staffordshire, Radnor and Montgomery.

In Wales the now extinct, tiny, tan-faced Rhiw and Cardy sheep were modified into the small, white Welsh sheep which nibble the mountain tops today. There is also a black Welsh Mountain sheep and a Tordhuu, which has similar facial markings to a bader. Further down the slopes an amalgam of Welsh Mountain and Border Leicester results in the Welsh halfbred and, in the west, round Anglesey, the handsome Lleyn sheep.

The free-ranging grazing habits of the Norfolk Horn sheep did not fit the new four-course systems of farming developed in East Anglia. In 1973 the last Norfolk ram was mated with Suffolk ewes and the New Norfolk Horn was born. It is ironical that the Suffolk was used to reinstate the breed as they had brought about the Norfolk's demise in the beginning. The Suffolk was the result of a cross between a Norfolk and a Southdown and it was tractable, had a better carcase, and matured quicker. From the Norfolk, the Suffolk gained its black face, a degree of hardiness and 'a good lean flesh' and it is now one of the most popular breeds in Britain.

In the north of England the New Leicester created three new breeds and influenced several others. It transformed the sheep of the Teeswater river valley from 'tall, clumsy animals, polled and with white face and legs' into ones of extraordinary fecundity who invariably produce twins. From the Teeswater evolved the Wensleydale. The Wensleydale was also used to give size to British hill breeds, the results emerging as the now famous half-breds, or first crosses, the Masham and Greyface.

The incentive provided by these experiments was adapted to improve other local sheep. The black-faced sheep of the Peak District of Derbyshire were refined into the Derbyshire Gritstone, and excessively hardy similar-looking ones inhabiting the Pennines in North Yorkshire were transformed into the Swaledale. Another horned sheep emerged, after a dash of Merino and Cheviot blood, as the Whitefaced Woodland.

Further north, on the Cheviot Hills, the Border Leicester continued to gather strength and popularity as a breed. By 1850 its half-bred progeny became the principal export of the border hills; the wethers were satisfactorily greedy and fattened speedily into meat while the ewes were excellent mothers who produced fine, early maturing lambs if given plump, downland-breed husbands.

Scottish landlords also took note of the success of the new sheep. Following the Jacobite Rebellion many of these chieftains, deprived of their old privileges and power, became absentee landlords only interested in rents. The shortages created by the Napoleonic Wars had raised the value of wool and mutton. However, the indigenous, fine-woolled, dun-faced sheep were ill-equipped to meet the demand. Small in stature and in numbers they were farmed by peasants whose methods in many cases had hardly advanced beyond the middle ages. The landlords attempted to replace the unprofitable indigenous sheep with more viable breeds. Frustrated in their attempts by a resisting human population they resorted to other methods. Raising rents, it was discovered, was a simple method of driving people to hunger and despair and forcing them, eventually, to move. Those who objected were evicted, frequently by the violent expedient of having their house burnt over their heads. For most emigration was the only solution: between 1780 and 1808 some 12,000 left the Highlands for America and at least 30,000 for the Colonies. By 1820 the depopulation of the Highlands was all but complete. The moors were turned into vast sheep walks.

To give the landlords some credit, the belief that sheep farming was the solution for the Highlands was often genuinely felt. What to do with those bleak, infertile wastes was an intractable matter. They were unsuitable for crops, and cattle had to be housed in winter, whereas sheep could live outside all year round. But the growth in the sheep population soon began to cause an imbalance in the native flora; they also drained the fertility of the land at a higher rate than cattle, so its sheep-carrying capacity has slowly diminished.

The breeds of sheep that caused most of the upset were the Blackface and Cheviot. The Blackface supplanted more people than the Cheviot. This rather dour sheep is popularly thought to have originated in the border forest of Ettrick. A flock of 5000 was planted at an unknown date by 'one of the Scottish kings' to supply the Royal household. Others ascribe their beginnings to the Pennine moors. From these 'black' hills the Blackface are thought to take their name and not from the colour of their faces which are actually as often speckled with white as wholly black. Wherever their home, their hardiness and ability to shift for themselves in an often inclement climate led to their widespread adoption and by 1941 they accounted for 75% of the 2,270,000 sheep in Scotland.

Fast on the heels of the Blackface came the Cheviot, a sheep with a high-held head, roman nose and arrogant air, sometimes called 'the long sheep' after the distance between eye and ear and lengthy body. The Cheviot, although not as hardy as the Blackface, has superior wool. Most are still found on the 'green' hills of Scotland, the lower, grassier hills of the Borders, Sutherland and Caithness.

Blackface and Cheviot, Merino and Leicester were introduced by landlords in an attempt to improve the native stock of the far north of Scotland and the Shetland islands which had become run down through mismanagement.

The new blood did improve the sheep but coarsened the Shetland's wool. Much damage was done before it was realized that the real value of the Shetland is its soft wool and ability to endure the inclemency of the climate and habitat. The true Shetland, told by its short flounder-fish-shaped tail, is now seldom found except on some of the remoter islands such as Foula and Papa Stour. A variety, formerly called The Orkney sheep, and found on all the northern islands, was rediscovered on North Ronaldsay and rescued from extinction in 1974. Here the tiny sheep looking more like over-grown lambs had, as Low graphically describes:

> acquired the characters which fit them for the conditions in which they are placed. The country which they inhabit possesses a climate eminently cold and humid and is exposed to continual gusts and storms . . . under these circumstances, the sheep are small in size, but hardy and capable of subsisting under great privations of food.

They also developed the peculiar attribute of being able to live on seaweed. Over the centuries the sheep's digestive systems became so adapted to seaweed that, given too much grass, they developed copper poisoning and died. Seaweed contains a copper-blocking agent and the sheep's bodies learned to assimilate copper four times as efficiently as other breeds.

The Western Isles also escaped the intrusion of foreign sheep until well into the 20th century. There the four-horned, black Hebridean and the delicate deer-like Soay continued to provide the background of the economy. The wool was carded and spun into yarn by women who, sitting together, talked and sang as they worked. The yarn was woven into tweed by the men, all expert weavers who worked their looms through the winter evenings, often into the early hours of the morning. These sessions were frequently joined by a fiddler and progressed into dances which sometimes finished so late that straw beds were made up in a barn for people from a distance. From this habit grew a custom known as 'bundling', where a courting couple were allowed to spend the night together in the girl's bed, but fully clothed to prevent anything untoward happening.

In these remote mountain areas methods of sheep management remained fairly primitive until the end of the 19th century. Wool was plucked from the sheep by hand, or with the aid of a penknife, centuries after shearing became commonplace in other parts. George Murray, an undergraduate from Aberdeen University who spent a year as a school teacher on St Kilda in 1886, records his efforts to introduce shears:

28 May 1887 – The sheep are to be clipped today. I clipped seven yesterday and learned them the way to use shears. Their shearing instrument is the common knife which of course makes work. One man last year got a pair of shears in a present and on my making use of it yesterday it became an object of wonder and was called a *great invention* . . . Remarks such as the following were made: 'O love don't cut the throat. Don't take out the liver.' While the owner of the beast said it would not stay on that side of the island after hearing such 'ghogadich' noise about its ears, meaning the sound of the shears.

The sheep were not rounded up but run down by dogs who were also trained to catch seabirds. The dogs, whose teeth were filed down regularly and their canines removed at six months, to prevent them harming the sheep, would catch them by the neck and hold on until relieved by a man; a hit-and-miss method that could result in both frightened sheep and over-enthusiastic dogs leaping to their deaths over clifftops.

The wild and primitive sheep is, as W. Youatt says, 'an intelligent and courageous animal, capable of escaping, by artifice and swiftness of flight, from his larger foes, or of beating off his smaller enemies by a dexterous use of the weapons that nature has given him', and rams have been known to attack dogs and foxes. Their domestic counterparts, in contrast, run at the merest hint of aggression, which Youatt puts down to a lack of education, saying that since 'our connection with sheep extends no further than driving him to and from his pasture, and that at the expense of much fright and occasional injury, and subjecting him to painful restraint and sad fright when we are depriving him of his fleece' his intellectual powers become depressed.

One of the few breeds in England to retain its natural self-reliance is the Herdwick. This large, grey-blue, teddy bear of a sheep, whose present looks owe something to the introduction of blood of other Pennine breeds in the 19th century, has grazed the windswept hills of Cumbria from an unknown date. They take their name from Herdwick the old north country name for sheep pasture and survive the harsh environment of the hills better than any other sheep. Their present prosperity owes much to the interest taken in them by Beatrix Potter. When she gave up writing children's stories, moved to the Lake District and took up farming, she adopted the Herdwick sheep. Her enthusiasm revived the breed which had waned along with the local tweed and knitting industry.

The first mention of knitting goods for sale in England was in 1488 when an Act of Parliament mentioned the price of 'knitte caps' as being 2s 8d. The knitting trade expanded and was encouraged by Elizabeth I as a method of poor relief. Knitting was taught in schools. In Wales, Scotland and the Lake District, Cottage industry grew up, suppliers taking weekly quantities of wool to the knitters, fetching the finished garments – mainly socks and caps – and selling them at market. At the end of the 18th century Bala 'great market' in Wales sold £2,500 worth of knitted goods each week, the work of 5000 people including wool-combers, spinners and knitters. Now knitting only remains a cottage industry in Shetland.

The middle years of the 19th century were prosperous for all. Britain had defeated Napoleon, Queen Victoria was on the throne, the new crops and animal breeds were being successfully produced and the new industries were at their height of production. This boom era, which lasted from 1837 to 1874, was equally good for sheep. Meat and wool were fetching high prices – Lincoln wool rose from 13d in 1851 to 27d in 1864 – and the new breeds were perfectly tailored to fit the market. Between 1866 and 1974 sheep increased in numbers to over 30 million and the average quality improved.

It must have seemed impossible for the good times ever to end, but their very success planted the seeds of their own downfall. The new Leicesters were exported abroad, and by the middle of the 19th century, the influence of 'the distinctive long white face, long alert ear and lively carriage of the Dishley blood' was to be seen throughout the world. The new Leicester reached North America in 1799 and was well established in Australia and New Zealand by the early years of the 19th century. The Corriedale, that great New Zealand wool and meat breed, bears its hallmark and on the continent its stamp is most obvious in the Dutch East Friesland breed. The colonists then began breeding their own animals with increasing success and to pay in kind for the English manufactured goods that they had yet to make for themselves. 'Food was, so to speak, the currency in which foreign nations paid for the English manufactured goods, and its cheapness was an undoubted blessing to the wage earning community' (Ernle). Peel's Free Trade Act of 1842 further increased the quantities of imported food and a better knowledge of food preservation accelerated the flow. Until 1877 cattle and sheep had been sent in live but after 1899 tinned and boiled meat enabled imports to be enlarged from 181,000 cwt in 1882 to 3½ million cwt

72

in 1899. Frozen meat then became feasible and Argentina and America joined the trade. Swamped by all this foreign competition and hampered by waves of disease, British sheep farming reeled.

The 'sunless ungenial summer of 1879 with icy rains' (Ernle) was the culmination of a series of bad summers, which resulted in an outbreak of pleuro-pneumonia, foot rot, liver rot and foot-and-mouth disease which killed millions of animals including sheep. Many of these diseases, including foot-and-mouth which became endemic in 1839, were new and derived from the importation of sick animals. The Contagious Diseases of Animals Act of 1896, which required all stock to be slaughtered at its port of landing, helped and so did the growing knowledge of veterinary science. Gradually the causes of sheep infections and infestations were identified and cures discovered. The double dipping of sheep in an arsenic bath controlled scab. Thomas, an Oxford zoologist, identified the mechanism of liver rot in 1881 while Simonds distinguished several types of lung worm in 1874 and showed the close association between pasture management and the degree of parasitic infection in sheep.

These controls had little effect on foreign competition which continued to expand. In the meantime the high fertility of British land began to decline and in 1893 a Royal Commission was appointed to enquire into the disastrous position of agriculture. It found that the value of produce had diminished by nearly a half while cost of production had risen. Several Acts were introduced, among them the Fertilizers and Feeding Stuffs Act of 1893 and Improvement of Land Act of 1899, which ameliorated conditions. Even so agriculture was never to be the same again. Neither were the farmers, although by the turn of the century they had become, according to Lord Ernle, 'for the most part alert, receptive of new ideas, keenly sensible of their debt to science, eager to accept its latest suggestions'.

The sheep were consolidated to thirty to forty recognized breeds. The Blackface dominated the high moors of Scotland, the Cheviot the lower, green hills and in the valleys and lowlands, the Border Leicester. Herdwicks prevailed in the Lake District, while black mottled-faced breeds divided up the rest of the north of England; there was the Rough Fell in Westmorland, the Lonk in north Lancashire, West Yorkshire and part of Derbyshire; the Swaledale inhabited the north Yorkshire dales and the Derbyshire Gritstone the southern ones, its territory spreading into Derbyshire usurping the land of the old Whitefaced Woodland and Limestone breeds. The short wools, the Welsh Mountain and Kerry, grazed the hills of Wales while the Ryeland, Radnor, Clun and Shropshire were distributed over the marches. In the south-west the long-woolled Exmoor Horn and the Devon Longwool were joined by a new breed created in 1923, the Devon Closewool. The Dorset Horn also had some fresh competition for its heathy hills from the Dorset Down, first registered in 1904. More down breeds were spread across the south of England, Hampshire, Oxford and South Down smoothed the surface of the downland of these counties. The Romney Marsh still grew fleece as 'white as a piece of writing paper' on the Romney Marsh and the Cotswold, Wiltshire and Leicester were still grazing their land. But the Norfolk's territory had been taken over by the Suffolk.

This pattern lasted until the First World War. The social upheavals caused by this bloodbath, the ensuing unemployment and recession all had their effect on the sheep world. Foreign competition exacerbated the situation, wool prices slumped as the markets were flooded with excellent cheap wool from abroad. Obeying the principles of Free Trade farmers concentrated on the two commodities least exposed to foreign pressure, fresh meat and milk. Smaller families became the norm and the demand came to be for smaller Sunday joints but, even when families ate a joint, they wanted different food on other days. The 1930s saw a massive recession in the sheep industry and thousands of sheep farmers went to the wall.

During the Second World War government policy favoured cattle over other livestock and much of the rough grazing ground of the downs was ploughed up to grow grain crops. This also cut at the base of sheep-keeping in Britain. As a result sheep, although still the second most important grazing animal in the country, only made up 21% of the total. There were also escalating labour costs and the continued growth of towns with their attendant populations of sheep-worrying pet dogs.

After the Second World War the neglect of wool in favour of mutton began to have an effect. The scarcity of quality wool sent prices rocketing and the Wool Marketing Board was founded to stabilize prices and set standards. All farmers with more than four adult sheep now have to sell their fleeces to the Board which sorts and grades them. They recognize more than 200 grades of wool categorized by length of staple – a piece of wool pulled from the fleece – its softness or harshness, colour, gloss or lustre, and coarseness.

Most British wool is sold at auctions held every two weeks at Bradford and Edinburgh and from there it makes its way, often through other auctions, to destinations all over the world. The coarsest frequently goes into carpets and Italian mattresses. Because of the heat Italian mattresses are often emptied and washed; coarse, strong fibres absorb perspiration and can be cleaned without losing their springiness. Fine wools are taken by cloth and woollen manufacturers.

The greatest number of sheep now live on the hills of Britain and it is here that most of the approximately seventeen million sheep in Britain exist. Although sheep prefer well-drained pastures on chalk and limestone, they are better able than any other stock to endure rigorous upland conditions. Sheep require little attention and no housing which 'maximizes national resources' by producing profitable results from poor land. On hill farms which vary from 100 to 3000 acres all the land will be devoted to pasture and, although a small herd of cows may be kept for breeding or beef, the main interest of the farm will be sheep. A 3000-acre farm will probably have about 1000 ewes, 100 hogs and something over 1000 lambs. Densities are highest in Wales, where the sheep are small and the grazing good, and lowest in the north of England where large-framed sheep have to feed off poor-quality herbage. In the rich grassland areas sheep are stocked at 2 to the acre and under carefully managed, modern intensive systems 6 to 9 are carried, but in the Highlands on the poorest ground, 1 sheep needs 3 acres.

The main aim of the hill farmer is the production of fat lambs for the meat market. Lambing begins in April and continues until the end of May depending on the altitude of the farm. In northern regions, where spring is late and feed is poor, ewes are only expected to produce one lamb in May whereas in kinder, southerly parts they can bear two lambs in April. The same breed has different lambing rates (expressed by farmers in percentages) according to its conditions. Blackface sheep have a lambing rate of only 65% or just over one lamb for every two ewes in North Inverness but one of 100% in Aberdeen. The ewes are brought to enclosed pastures round the farm for lambing. After two to three weeks ewes and lambs are put back on the hill, first on enclosed foothill pastures and later onto the unfenced hill. Before being turned loose, the sheep are ear-marked and their fleeces daubed with the owner's mark. At the end of June or the beginning of July the flock is brought down from the mountain for clipping. The sheep are then put back on the hill until around August 20th when they are fetched down again for compulsory dipping against parasites. This process is repeated in late October or early November. The rams are put with the ewes in about November. They stay up on the hill with the ewes for about three weeks before being brought back to the farm. The ewes remain on the hill all winter. If snow threatens, they are gathered into sheltered places to prevent them being buried in drifts.

Lambs from hill farms are generally sold in autumn, frequently to lowland farmers who finish their fattening. Female lambs required as replacements in the flock are often put on a lowland farm for their first winter to gain strength. Breeding ewes do not as a rule spend more than three or four seasons on the hill. After this they are often sold on to lowland farmers who breed from them for one or two years before slaughter. This system of 'flying' or non-permanent flocks whereby sheep are moved, generally in an eastwards direction, downhill underlies the whole structure of sheep farming.

The kind of sheep each farmer keeps is also carefully selected. Usually his choice is limited by his area and often by fashion. As Ralph Whitlock succinctly puts it, 'There are, for instance, no fewer than five distinct breeds of sheep in Devon, but let no amateur think that he has the choice of these five when stocking his farm. No, he must use the breed best suited to his district, or take the trouble which will inevitably follow. Besides this, there are certain well-tried and well-known crosses which he may use with safety. The Border Leicester × Cheviot, known as the half-bred, is a popular one for store lambs, and a Suffolk ram on half-bred or other first cross ewe usually yields good results.'

The main supply of British meat comes from cross-breds rather than pure-bred sheep, although, because cross-breds do not breed true, reserves of pure-breds are always required. Britain is unique in Europe, and probably the world, in her enthusiasm for cross-breds. The search for the perfect variety goes on all the time, breeders searching everywhere for new blood. The bulky Colbred is a result of an importation of Friesland ewes native to the Dutch-German border and noted for their prolificacy and high milk yields. These were crossed on Dorset Horn, Border Leicester and Clun Forest with the objective of breeding a sheep that would sire prolific, milky cross-bred sheep out of native hill ewes. Finnish Landrace sheep were imported to transmit their high sexuality to British breeds but their disadvantage is that they need housing in winter. The latest arrival from abroad is the French Texel, a lean sheep which transmits these qualities to its progeny making them suitable for the modern meat trade. This meat trade is now introducing new grading standards to encourage the production of lean lamb for modern palates which dislike the taste of fat and believe it causes human disease. The British, nonetheless, consume enormous quantities of sheep meat. In 1979 Britain produced 237,000 tonnes of sheep-meat, almost half the entire EEC production, and she was also the largest importer of sheep meat.

In spite of the cross-breds, it is trusted favourites like the Suffolk and the Blackface that most consistently break cash barriers in the auction ring. Prices for good rams are frequently in excess of £1000 and recently £11,000 was paid for a Blackface tup. Rams are still needed in sheep farming, for AI, which predominates in the cow world, is not generally practised on sheep. Sheep meat is the purest meat available, freer than any other of additives and chemicals, but the sheep themselves are hardly the rugged individualists of former times. John Stewart Collis in his book *The Worm Forgives the Plough* deplores their

> lack of self-possession and confidence. Their perpetual hurrying and scurrying, their weak faces, their ceaseless maa-ing and baa-ing make one feel somehow they have got lost in evolution and are in a frightful state of anxiety about it. Yet it is no wonder they feel bewildered and lost. Like so many creatures, they are now half created by Nature and half by man. The result is that they are now dependent upon us for everything.

On many farms they are rounded up by shepherds who have abandoned walking for tractor bikes. Diseases are cured by a bewildering array of dips and drenches. Some practices have come full circle. The *Farmer's Weekly* publishes articles on the latest developments in sheep farming, some of which have been heard of before. The housing of sheep in winter, common practice in the middle ages, is put forward as 'one way of intensifying production without using more land or labour'.

Milking sheep is another new, old idea. It is suggested for third world countries unsuitable for cows. So keeping high-milking, prolific, triplet-producing ewes is a way of contributing to farming profits in the 1980s and milk sheep are making headway. But disturbingly 'lamb for veal' is being put forward as a way of dealing with the multiple offspring of sheep who have not enough teats or milk to go round; the lambs are reared on milk replacer and slaughtered at five weeks old. The disadvantage of all these intensive methods of farming is the high level of stockmanship required to maintain output, and it may be that other older, more self-sufficient breeds who make use of marginal land, require little nourishment and produce a lean carcase will win the day. But perhaps there is no end to the story and sheep will continue to evolve as they have since domestication began, their shape partly dictated by man's needs but also dictated by their own natural laws and their landscape.

The Cow

The auroch is the ancestor of all domestic cattle. A huge, thick-set animal with spreading horns and wild eyes standing up to six and a half feet at the shoulder and of enormous strength and fierceness. Its bones are found in glacial deposits from the time of the Upper Pliocene. They are most common in a circle at the top of Arabia encompassed by the Persian Gulf, Caspian, Black and Mediterranean seas. These bones date from about 6000 BC and it is then that domestication is thought to have begun. The process of acquainting auroch with man probably began by placing salt and water regularly in places near human settlements. By 4000 BC, it is assessed, the auroch had become an established farm animal and by 2900 BC, judging from some drawings on cylinder seals found in Mesopotamia which show cattle being fed, bulls being driven, drawing a sledge and carrying a shrine, and a cow being milked, they were well in hand and developing into different types.

Why did man attempt the improbable task of taming the auroch. The reason could not just have been their meat which could as easily have been got by hunting, or their skins which early man used for shoes and shields, or their dung valuable even then as fertilizer. It may have been for their milk. Early man was familiar with milk from his long association with goats. But the over-riding reason, it is thought, was to use cattle as beasts of burden and traction. The development of wheeled vehicles is closely connected with the domestication of cattle, and cows are still the main pullers of carts and bearers of men in most primitive parts of the world.

By 4000 BC the domesticated cow was spreading into Africa and improved types were being developed. An Egyptian mural relief of 2,500 BC shows not only long-horned cattle, direct descendants of the auroch, but polled or hornless cattle and short-horned ones as well as some with pied coats and large udders. Their daily functions are depicted in small models, sculptures and painted reliefs. There are cows drawing ploughs, cows being milked and cows with calves. Bulls are shown being driven on threshing floors trampling the grain, being roped, thrown and slaughtered. Reliefs depict cattle grazing and being moved about the countryside. They are also drawn being hunted. From early times cattle were also objects of myth and worship. Primitive man believed that earthquakes happened when the monstrous bull who held the world between his horns shook his head in anger. The palace entrances of the Assyrian kings were guarded by winged bulls with human heads and their furniture was decorated with bulls, cows and calves in finely carved ivory. It was in Egypt, however, that the height of cattle adulation was reached. Here sacred bulls noted for their wisdom and prophetic gifts lived in luxury and were even provided with harems of cows. The most famous of these sacred bulls was Apis, a black bull considered to be the reincarnation of the god Ptah and his son by a virgin heifer. A cow, Hathor, was the protector of women and also a goddess of love and drunkenness who sanctified the deeds of the Pharaohs.

The cult of bull worship reached its most complex during the Minoan period in Crete. The skulls of very large bulls adorned the walls of the palace of Knossos and, it is thought, they were regularly sacrificed in special rooms perhaps marking the beginnings of bull-fighting. The inter-mixing of bull sacrifice, worship and sport culminated in the spectacular bull-leaping acrobatics depicted in the murals at the palace. The game was played before large audiences in an arena and the bulls were confronted by teams of unarmed men and women who grabbed the horns of the charging bull and somersaulted over his back.

Bulls also entered Greek mythology. Zeus, ruler of Olympus, turned himself into a bull to gain ascendancy over the hesitant Demeter. Pasiphaë's infatuation with a bull led her to hide in a specially made dummy cow. Her seductive ruse succeeded, bestiality was performed, and the unnatural union resulted in the Minotaur, half human, half bull, a monster condemned to wander the labyrinth under Knossos for life, his craving for human flesh satisfied by an annual tribute of seven Athenian boys and girls until he was slain by Theseus.

The most bizarre form of bull worship was the cult of Mithraism. Although there are no written records of this all male freemasonry sworn to secrecy, Mithraism was the most serious rival to Christianity in the 3rd century AD.

Early man admired the bodily strength and sexual vigour of bulls partly because he required both these qualities to combat the many extinguishing elements of his existence. Although the Assyrians, Egyptians and Greeks had reached heights of civilization, for those outside their world life was extremely barbaric. The nomad pushed through the forest in search of game, creating little settlements here and there and establishing corrals of stock. He wandered rather than journeyed, only making a move to the next place when pressed by warring tribes or land exhaustion. Thus, in galvanized stages, he progressed out of the Middle East and extended his perimeter, east, west, north and south. Nomadic man's cow was still basically a long-horn but new characteristics began to form in response to each country's climate and terrain. Most noticeable are the muscular humps and pendulous dewlaps that developed on the cattle which entered the hot, dry climates of India and Africa. A cow's hump is thought to perform much the same function as a camel's, acting as a store for water, while dewlaps give an added area for the evaporation of sweat and keep the animal cool. Both these features can be seen in the Indian Zebu with its long lugubrious face and steep, upright horns, the African Kuri cattle of Lake Chad and also the Saiga, Fulani, Nguni and Africana breeds.

When Neolithic man reached Europe with his stock he found the wild auroch already well established. As the ice age retreated and vegetation grew, the auroch, naturally a beast of open forest and tall scrub gaining much of its food from the leaves of trees, spread gradually over Europe.

The wild auroch also crossed into Britain, probably when it was still joined to Europe. By the time Neolithic man attempted the journey the English Channel had created a watery gap. This he crossed in about 2500 BC and, in all likelihood he brought his own cattle, dark shaggy-haired, horned beasts with wide foreheads resembling Highland cows. These cattle were still probably half-wild, to be handled with circumspection and as seldom as possible. The cattle were kept in herds, rounded up in autumn into earthen-banked kraals; the causeway of Trundle and Whitehaven in Sussex, Windmill Hill and Maiden Castle in Dorset are all remains of pounds used for round-ups when cattle were marked and some selected for slaughter and breeding. The rest of the year, it is thought, the cattle roamed free fending for themselves off grass and other vegetation. When an area was exhausted they were driven to new pastures down routes along the tops of hills which avoided the dense forests and swamps of the lower valleys. These tracks became established and form the earliest drove roads. The Icknield Way which stretches from Dorchester to near Norwich was probably a neolithic drove road, its name derives from the old word for oxen 'yken' or 'ychen' and its whole name means 'the way the cattle go'.

As domestication increased, the size of the cattle decreased. Remains at Maiden Castle in Dorset reveal a reduction in horn length which archaeologists attribute to a policy of killing the largest calves for meat and retaining for breeding the smaller beasts which would have been easier to handle.

With time man became more settled, cleared more forest which gave grazing for that entirely pastoral animal the sheep, and the numbers of cattle declined. Then, between 1900 and 450 BC a people called the Beaker Folk arrived, breakaways from the Swiss Lake dwellers, who had achieved a relatively high standard of living for the time. The Beaker Folk brought with them a new kind of cow; small, slender and narrow with short horns curving forwards and generally called *Bos tauros* from the length of its forehead. Under their care it had become a milk cow and it is to this Celtic cow that most European dairy breeds owe their origin. The Celtic cow's hair was red, anyway in Scotland. The evidence for this are some hairs found in bog butter dug up in Scotland. These cows were housed in cow stalls for the first time; tethered to the wall at Jarlshof in the Shetlands by a ring made from a whale's vertebrae. The house had a sloping floor and a central collecting tank to gather manure used as fertilizer to spread on fields, the first of which date from this time. Square plots dug with the innovatory plough on the 'cross-plough' system. These bronze ploughs, unlike a modern plough whose curved blade turns the sod over, merely dug straight into the ground. To turn the sod fully the land had to be ploughed again at right angles which made a check pattern. Pulling this new plough was a new animal, the ox. It is said that 'The first man who made his cow instead of his wife draw the plough was a great benefactor' and the plough oxen was, from now on, to be central to British farmyard economy until displaced by the horse.

An ox is a castrated bull. Its testicles are removed or destroyed by what must be one of the oldest surgical procedures in history. Ever since domestication began the bull has had to suffer the pain and indignity of castration. Originally scorpions were used, later crushing stones, then a knife. Now, most commonly, tough rubber rings are fitted round the testicles to compress their blood and nerve supply until they rot and fall away, although castration by injection has recently been pioneered by an Edinburgh-based team of researchers. Permanently sterile and impotent, the ox's secondary sexual characteristics, those which separate male from female, never properly develop, its body structure changes, it grows fat, loses all its male aggressiveness, becomes docile and tractable and infinitely more trustworthy.

In late Celtic times the normal ox team consisted of four oxen abreast attached to a twelve foot long yoke and guided by two men; a ploughman whose job was to guide the plough from behind and a driver who had charge of yoking the oxen 'so that thye be not too tight nor too loose'. Ploughing started legally on February 9th and oxen were expected to work from morning to evening for their first two years but allowed off at midday in their third and succeeding years. The amount of land an ox was expected to cover in a day determines the measurement of many of our fields and distances. The Celtic field was divided into long narrow strips called 'erws' separated by a bank of turf two furrows wide. The 'erws' represented the usual amount ploughed by an eight ox team in a day while the length of the 'erws' was a 'furlong', or 'furrow long', and equalled an eighth of a mile or the distance an eight ox team could draw a plough without stopping for breath as primitive harness pressed on the windpipe constricting breathing.

Pulling these ploughs were small dark cattle and from these all the black breeds, the Galloway, Aberdeen Angus, Kerry, Dexter and Welsh Black, are said to descend. The legal Celtic herd consisted of twenty-four cows and one bull and was made up of cattle from the various households of the hamlet and tended by a single herdsman. In winter the people and cattle lived in the valleys but in May all except probably the plough oxen, moved up the mountains for the cattle to gain the benefit of the hill pasture. Dorothy Hartley in *Food in England* admirably describes how

The herds came up from the valley and were milked on the mountain pastures all through the bearing season. As the flow of milk increased, workers, idle (after spring sowing) on the farm below, came up and joined the mountain dwellers in making cheese and churning, washing and salting down of butter. They harnessed the water-power of some mountain stream, and used it also for the washing of troughs, etc. Probably the distilling of a whey spirit was general.

... The mountain workers lived on ... oatbread and butter and cheese and curd dishes all summer, learning the variety of mountain herbs that they watched the beasts relishing ... [In the evenings] they sat around in the firelight and spun wool and sang; while the wind howled through the rocks above and the stream piped below.

In Celtic times cattle were valued not only for their milk, flesh, hides as well as for their tractableness but, until the establishment of coinage, they stood as a standard of value. A tribesman's wealth was in his herd and the value of everything was expressed in terms of so many cows. The Belgae, a rich and sophisticated people, altered this system; they introduced coinage and, being skilled stockraisers and dairy farmers, kept many cows. Although most of these were based on the stocky, black, dark brown or dun coloured Celtic cow, some were of a different type, a fact deduced from their custom of casting the bronze handle mounts of cauldrons and buckets in the shape of cattle heads. These fall into four broad types; most common are those with horns going round, up and out, like an Ayrshire cow and others pointing forwards like a Jersey; rarer are those of the longhorn and little Celtic shorthorn.

Caesar tells us that the Britons 'neglected tillage and lived on milk and flesh'. The Romans only extensively settled and cultivated the southern part of England and probably introduced their own breeds of cattle. Here soils are lighter and a simple plough, which needed only two oxen to draw it not four, was suitable for most requirements. But two oxen were not able to plough as long a furrow without stopping for breath, consequently the Roman furrow was shorter than the Celtic, measured merely 120 feet, and the amount of land a team could plough in a day was short and broad and known as an acre.

The measurements of all early 'cruck' houses for humans and oxen were also dictated by the ox-team. Four oxen were allowed sixteen feet of standing room and this became the recognized width of the 'bay', or space, between each of the timber crucks used in constructing these buildings until the 16th century. Cattle houses were restricted to villas and the more civilized parts, on the downs and uplands cattle were driven into stone or earthen enclosures at night to protect them from the ravages of wolves.

Shoes for the soft hooves of draught oxen were also a Roman introduction, made necessary by their paved roads. Known as 'cues' or 'kews' ox shoes are thought to have been invented in Gaul. They are usually lighter than horse-shoes and comma-shaped. One was fitted to each side of the ox's cloven hoof; although sometimes only the outer clove was shod. Shoeing an ox is a tricky business, first the feet of the ox have to be tied together and the beast thrown on its side, then the shoe is attached by nails. Because of the thinness of the hoof, smiths are 'apt to drive the nail into the quick, and from the brittleness of it, it will not hold the nail'. At a later date shoes that covered the whole hoof, and turned up between the toes, were occasionally used and farmers near bogs, especially those bordering the Chet Moss bog, shod their cattle with wooden pattens in case they should stray.

The average Roman herd consisted of fifteen cows and one bull. The cattle were growing larger and had larger horns. Excavations at Woodyates reveal a beast standing four feet two inches at the shoulder (although the average was shorter), whose horns curved broadly. Palladius recommended black cows and red or brown oxen and these larger cattle came to dominate over all others.

White cattle were also liked by the Romans. Theories abound as to when white cattle came to Britain. Some say they were always here. Almost all the existing herds are established in areas of ancient British kingdoms and the ferocious Brigantes of the north of England, who offered such resistance to the Roman invaders, are said to have possessed herds of white cattle. The druids, when they ran short of human sacrificial offerings, resorted to white bulls. The Romans also sacrificed cattle, reserving white ones for the 'celestial deities' and black ones for the 'infernal deities'.

It is interesting that even after the arrival of Christianity the sacrifice of bulls did not cease and continued in remote areas until comparatively recent times. In A. Mitchell's strange book *The Past and Present* written in 1880, he relates how St Aelred of Rievaulx on a journey to Pictland in 1164 'happened to be at "Cuthbrichtis Kirche", or Kirkcudbright, as it is now called, on the feast day of its great patron. A bull, the marvel of the parish for its strength and ferocity, was dragged to the church, bound with cords, to be offered as an alms and oblation to St Cuthbert.' In 1767 a heifer was burnt alive in Mull and in 1850 an ox was sacrificed to drive murrain from a herd in Dallas, Moray, while in the 19th century a cow in Orkney was roasted alive by a farmer in a kiln 'as an offering to the disease spirit'.

In the troubled times that followed the departure of the Romans their herds of sacrificial animals were released into the forest. Here they would undoubtedly have mingled with local coloured breeds. Owing to the dominance of the white-coloured gene their characteristic colour was preserved but they gained black points and scattered markings. Even today White Park cattle occasionally throw black calves and all the herds vary in conformation. As time went on and the forests were thinned and contracted by felling these ferocious cattle became dangerous to the inhabitants of the neighbouring districts. In 1225 Henry III removed many forest areas from Royal Protection and the resulting free-for-all endangered the existence of many animals. Nobles and clergy, fearful of being deprived of their favourite pastime of hunting, formed parks into which were driven beasts of the chase including cattle. The white herds of Chartley, Chillingham, Needwood and Cadzoe were all imparked at this time.

Hunting cattle continued to be a recognized sport up to the 19th century. According to Mr Culley writing in the 18th century of Chillingham cattle,

> on notice being given that a wild bull would be killed . . . the inhabitants of the neighbourhood came mounted and armed with guns, etc., sometimes to the amount of an hundred horse, and four or five hundred foot, who stood upon walls, or got into trees, while the horsemen rode off the bull from the rest of the herd until he stood at bay, when a marksman dismounted and shot.

Wild cattle of all types abounded in the stretches of moor, bog and primaeval forest which separated villages until quite late on. There were bulls in the Chiltern forests of Buckingham, Hertfordshire and Middlesex, and Enfield Chase, which extended to the gates of London, held 'wild bulls in abundance'. Across the border in Scotland white bulls 'with crisp and curling manes, like fierce lions' roamed the Caledonian forests until the 16th century and Hector Boece ghoulishly describes in his *Scotorum Historiae* how

> thoucht thay semit meik and tame in the remanent figure of thair body's thay wer mair wild than ony uthir beistis, and had sic hatrent aganis the societi and cumpany of men, that thay come nevir in the woddis na lesuris quhair they fand ony feit or haund thairof: and, mony dayis after thay eit nocht of the herbis that wer twichit or handillitt be men.

Of all the fourteen herds of White Park cattle that now exist the best known is the Chillingham. They are also the purest, all the others have, at some time, been infiltrated by foreign blood. Scientifically Chillingham cattle are interesting for the fact that they have survived centuries of in-breeding. The Earl Tankeville attributes 'Their remarkable survival' to the

> fact that the fittest and strongest bull becomes 'king' and leader of the herd. He remains king for just as long as no other bull can successfully challenge him in combat, and, during his term of kingship he will allow no other bull to mate with any of the cows. Nature seems thus to have ensured the carrying forward of only the best blood. (*The Field*, 16th October, 1948.)

In this and many other habits White Park cattle resemble wild cattle. The cows hide their calves in undergrowth for a week to ten days going to suckle them several times daily. At the threat of danger the calves 'clap their heads close to the ground, and lie like a hare in form to hide themselves' and the whole herd comes to their rescue the dam attacking the intruder 'with impetuous ferocity'. If any cow 'happens to be wounded, or is grown weak and feeble through age or sickness, the rest of the herd set on it and gore it to death'. The cattle have no rutting season, calves are born throughout the year, and males and females are born in equal numbers.

As Earl Tankeville described, the strongest bull takes all the females and the weaker or younger bulls are forced to live monastic lives. At intervals a younger bull of exceptional strength will challenge the king's supremacy. Bellowing at each other, the bulls paw the ground, rub the sides of their face and horns along the ground and go at each other butting their flanks with head and horns. 'Once one bull has established his superiority the conflict ends. The old king bull is not dead, but the new king reigns', says Fraser in his book, *The Bull*. Territorial rights are established by the bull pawing and horning patches of earth bare throughout his area.

The beast described in an ancient runic poem probably fitted the wild cattle rampaging the country when the Angles and Jutes arrived in the 6th century.

Ur is proud and crowned with horns
A very terrible beast, it fences with its horns;
It strides mightily across the moors and is full of courage.

The Angles, perhaps having heard of these ferocious beasts, decided to bring their own tamer cattle, packing them into their long boats along with other stock. Some of these cattle were red and it is thought that the brick-coloured breeds of the east coast, the Lincoln, Red Poll and Sussex are all direct descendants. Others, it is surmised, were polled and the dirty yellow-coloured Suffolk Dun, Yorkshire and Durham cattle owe their hornlessness to these cattle. Yet others were white and to these the British White, a polled breed with similar markings to the White Park, almost definitely owes its origin. British Whites were all originally found down the east coast, principally in Lancashire and later in East Anglia. The latter is still their main home except for a few recently exported to the USA.

The advantage of polled cattle over horned has been endlessly debated. Some argue that they eat more and so will fatten faster or give more milk but, in the end, the real asset of polled cattle seems to be that they can be packed closer together in pens and houses and, without their natural method of defence, are more docile and easily handled. Horns are an elongation of the skull's frontal bone. Unexpectedly they are not made of dead material but, rather like a horse's hoof, composed of a hard gelatinous material which covers the most vascular bone in the cow's body. Dehorning a cow produces a gush of blood equal to the amputation of a limb. The first polled cattle were probably accidental mutations as polled cattle do occasionally produce a freak horn which grows from the skin and, unattached to any bone, hangs down the side of the face. This is perhaps how the first polled cattle looked and from this they were selectively bred.

The growth of horns is closely related to sex but in cows, unlike deer where entire males have huge sets of branches and the castrated male grows smaller antlers or loses them altogether, the bull has short, straight, comparatively ugly horns while the ox has long, beautifully curved ones and cows horns of even greater delicacy. In South Africa, where they were ridden by ladies, cows with beautiful horns were prized, and those of young cattle were cruelly twisted into spiral curves, and other extreme forms, with heated irons. Horns have always had their uses. Youatt describes some and their treatment. After the outer horny case is separated from its core it is cut into three pieces and

1st. The lowest of these, next to the root of the horn, after undergoing several processes by which it is rendered flat, is made into combs. 2nd. The middle of the horn, after being flattened by heat, and its transparency improved by oil, is split into thin layers, and forms a substitute for glass in lanterns of the commonest kind. 3rd. The tip of the horns is used by the makers of knife-handles and the tops of whips, and for similar purposes. 4th. The interior, or cone, of the horn, is boiled down in water. A large quantity of fat rises to the surface, which is put aside, and sold to the makers of yellow soap. 5th. The liquid itself is used as a kind of glue, and is purchased by the cloth-dresser for stiffening. 6th. The bony substance which remains behind, is ground down and sold to farmers for manure.

The Norsemen, who settled the northern parts of Scotland and Shetland in the 8th and 9th centuries, brought a horned cow with them which is still sometimes found in the Shetland islands. It is small, with short slender horns and a sleek coat which grows as long as five inches in winter to protect the animal against the bitter cold. The Shetland became integral to crofting life. It pulled the ploughs, provided milk during its long life and, when dead, meat and shoe leather. Shetland cows are all black and white today, the result of an addition of Friesian blood in the 19th century, but formerly they were all colours. Each colour variant was given a name such as 'sholmit' or helmet for a white face. Their noises also had special names and merely by listening to their animals people could diagnose their state. Brül was normal, Gulbrül described an excited roar.

During the summer the young cows roamed loose over the 'scattald' or common grazing grounds, while the milking cows were tethered by their horns. In winter all were housed, fed a sparse diet of hay, straw, cabbage and latterly turnips, and stood on a deep litter of peat mould, dried grass and heather. By spring the byre was so deep there was hardly room for the cattle to stand but the material made rich manure. However, standing space was hardly required for the animals had become skeletons from their starvation diet and, unable to support themselves or walk, they had to be carried out each day in spring to graze the new grass until, by about May, they had regained their strength. This state was termed being 'in lifting' and although it sounds barbaric, the starvation process resulted in excellent beef.

In the south of England, in Saxon times, Alfred laid down laws regulating the behaviour of cattle and their owners. If an ox gored a man or a woman 'so that they die let it be stoned and let not its flesh be eaten' but if they

24. The Young Bull, P. Potter, 1647. Imported Dutch cows revolutionized our breeds.

25. Shorthorn bull, 1815. An example of the most influential 18th century breed.

26. *This engraving shows the disturbances caused by droves passing through London.*
27. *Smithfield market, 1858, before it was covered over, becoming a dead meat market.*

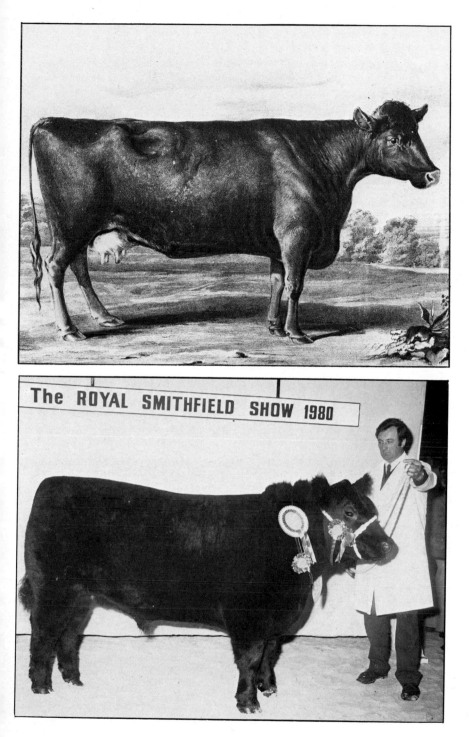

The ROYAL SMITHFIELD SHOW 1980

28. *The Suffolk Dun, now extinct, was a great dairy cow (18th–19th centuries).*
29. *Champion Aberdeen Angus,* Panda of Westdrams, *typifies the best kind of beef steer.*

30. A Charolais bull, a currently fashionable fast-fattening Continental breed.

31. Ten week old Friesian veal calves being fattened in crates.

32. *Przewalski's horses, the sole surviving type of truly wild horse.*

33. *Exmoor ponies, descendants of the original breed which pulled Boadicea's chariot.*

34. A horse and ox harnessed in tandem, a common habit which persisted in remote and heavy clay areas until quite recent times.
35. Norman soldiers at Hastings in 1066. To this victory the British owe the beginnings of their breeds of heavy horse.
36. Oliver Cromwell on his charger. Cromwell's victory over the Royalists was partly due to the fact that he drew his forces from the Midlands and East Anglia which traditionally bred the finest heavy horses. However the Civil War was the last time they were used in battle.
37. An Old English Black Horse, the precursor of the Shire.

THE
Right Honorable
and Vndaunted Warrior
OLIVER CROMWELL
Lo: Governour of
IRELAND

38. *Suffolk stallions at Ipswich Show in 1949. A breed prized for its pulling power.*
39. *Clydesdales for sale at Perth Auction in 1940: Scotland's only heavy horse breed.*

were merely injured the owner had to give them the cow. However, if the ox had warned of trouble by being 'wont to push with its horns' for several days before and the owner had not taken proper precautions he was liable to be slain himself. A retinue of hard-working, harshly treated people looked after the beasts. The Ploughman who tells of his day in the Anglo-Saxon *Colloquium* of Archbishop Aelfric says:

> O my lord, I labour hard: I go out at daybreak in order to drive the oxen to the field, and I yoke them to the plough. There is not so stark a winter that I dare stay at home, for fear of my lord, but having yoked the oxen and fastened the share and coulter to the plough every day I have to plough a full acre or more . . . I have a certain boy driving the oxen with a goad who also is hoarse from cold and shouting . . . I have to fill the stalls of the oxen with hay and water them and carry out their litter . . . Indeed it is great toil, because I am not free.

The ox-herd had the miserable task of guarding them at night 'to watch against thieves' and 'first thing in the morning' he had to 'commit them to the ploughman well fed and watered'. As recompense he was allowed to pasture at least two oxen and one cow with his lord's herd and received his keep, shoes and gloves. The cow-herd drove the cows to pasture, milked them and was allowed the milk of old cows after they had calved and the 'beastings', or the 'first milk', from any young cows for fourteen days after they had calved. The dairy maid made butter and cheese and part of her duty was to provide the lord's buttery with three sizes of cheese; great, middle and small varying in weight between two and six pounds.

Cattle were valuable assets and common grazing on unfenced woodland pastures meant that losses through negligence and theft were frequent. The stockman's recourse was to raise a Hue and Cry and the London Guild, in Athelstan's reign, awarded compensation for stolen cattle; a mancus was given for an ox and 20d for a cow. However, no one was trusted and they sternly announced that

> each man is to note when he has his cattle and when he has not, with his neighbour's witness, and if he cannot find it, to point out the trail to use within three days, for we believe that many heedless men do not care how their cattle wander, out of over-confidence in the peace . . . for we will not pay for stray cattle.

Land allowance was allocated according to the number of oxen a man owned. The possessor of a full plough team of eight oxen was allowed a hide, or 120 acres of land, divided into 30 acre portions; a gebur with two oxen got a yardland or 30 acres while a man who could only contribute one ox to the village plough team got 15 acres or a bovate.

The Normans perpetuated this system. In 1086 when the Domesday Book was compiled the team of eight oxen was intimately related to the measurement of land. In the village of Chenmere it is quoted 'there is land for 25 plough teams; and in the demesne (lord's land) there are two teams,

and there are 36 villeins and 11 borders with 17 teams (between them).'
Although eight was the standard team, numbers did vary. In the wilder area
of Norfolk two freewomen ploughed with two oxen while at Yelverton a
freeman ploughed with three oxen. At Kellaton in Devon the villeins used
two plough oxen and one ferling or pony, indicating that horses were coming
in and, as the 12th century advanced, mentions of their inclusion in ox
teams increase.

Plough oxen could either be steers or cows and generally entered the
team at four years old and, allowing for disease, calving and other lost time,
could spend from six to eight years in harness before being fattened for
slaughter. Their replacements came from village stock whose management
was dictated by the calendar. The cows were milked from May to
Michaelmas. At Martinmas, 11th November, the herd was inspected and as
many retained as there was winter fodder for, including the draught oxen
and breeding stock. The rest, 'the old cows with bad teeth, and the barren,
and the draft or the young avers that do not grow well' were sold or killed
and salted for winter meat. From St Luke's Day, October 18th to Holy
Cross, May 3rd, the cattle were stall fed, mainly on straw or hay, and curried
twice a day 'that they might lick themselves more efficiently'. After May 3rd
they were collectively driven out each morning by the village herdsman to
graze on common land. On Lammas Day, August 1st, the fences round the
village hay meadows were removed and the cattle allowed to wander at will
feeding on the aftermath and, in the process, manuring the ground for next
year's crop. These fields were accordingly called 'Lammas lands' and the
complicated grazing rights attached to them had a drastic effect on how
towns expanded in the 19th century. W. G. Hoskins tells how in his book
The Making of the English Landscape:

> Nottingham failed to solve the problem until too late and created as a
> consequence some of the worst slums in any town in England. Leicester
> solved it just in time and produced a town that spilled widely across the
> surrounding fields and gave its working class bigger and better houses,
> and wider streets, than almost anywhere in industrial England.

Fodder and its lack continued to bedevil the mediaeval cow who only
managed to produce 150 gallons of milk a year, whereas today a scrub cow
is expected to yield 300 gallons and a commercial milker 1000 gallons. But
milk was not the prime consideration of the mediaeval farmer; he regarded
his cow primarily as a source of plough oxen and these were given extra
food in the form of oats. Henley advocated that 'he should have at least
three sheaves and a half of oats in a week' and was the first to relate the
output of a cow to its nutritional intake.

Feed is also closely related to fecundity. Calving averages on many
mediaeval farms are high which suggests that on some farms keep was
adequate; on a farm at Wellingborough sixteen cows produced twelve calves
in 1258–9. However, on other farms averages were lower; on one at
Stevenage seven cows produced only two calves. In this case the bull could

have been at fault or its environment. Bulls and cows are as susceptible as humans to sexual stimulation. Some bulls prefer to mate out-of-doors while others show a preference for cows of a certain colour. Some females are definitely more attractive to bulls than others and, if a bull is running loose with a herd, he will choose these sexier cows first providing they are in heat. Bulls do not waste their seed on mere sexual gratification. A cow coming on heat bellows a lot, licks and grooms other cows and provocatively mounts them until her heat reaches its height when she stands and allows others to mount her. This period lasts from twelve to twenty-four hours. During this warm-up time the bull has not been standing idly by but has helped peak her desire by nudging and nuzzling the cow and licking her genitals until she is ready when, bellowing, he mounts her rump, tightly clasps her flanks with his forelegs, increasing the intensity of their pressure as, with a single pelvic thrust, he penetrates her vagina and ejaculates. His climax complete, he slips exhausted from her back and the two animals stand quietly head to tail grooming and licking each other over neck and shoulder, udder and scrotum until, with his energy and desire renewed, the bull mounts her again, and then again, five or six times in all during her heat period. Once this has passed he ficklely moves on to the next in his harem.

A bull's sexual energy is no greater than other domestic animals yet he has always represented to pastoral people 'the most natural type of vigorous reproductive energy'. In the Indian hymns of Rigveda the god of fertility, Parjanya, is compared to a bull who 'loud roaring, swift to send his bounty, lays in the plant the seed for germination'. It must have been the bull's strength and power which became confused in primitive man's mind with survival and procreativity. 'Power, irresistibility, brute force, these were the bull's qualities that made the greatest appeal for the people who thought of their king as god' said J. R. Conrad in *The Horn and the Sword*.

These heroic descriptions hardly fitted the mediaeval bull if accounts of his physique are to be believed. The author of the *Seneschaucei* might advise the use of 'fine bulls and large and of good breed' but, by and large, cattle in the middle ages were only a third of their present size and weight. They continued to deteriorate along with the land whose arable area grew less productive and infested with 'murrain'. The Black Death took its toll of the human population, decimating those available to toil on tenant farms, and wolves and the murrain took their toll of the cattle. Little is known of what form the murrain took, as the term was applied to every loss except theft and slaughter, but records suggest that undulant fever or brucellosis were already prevalent and could have been the cause of the losses.

Cattle continued to deteriorate until the Cistercians arrived in the middle of the 12th century. It is thought that these great livestock farmers and improvers, were appalled by the small, shaggy, disease-ridden indigenous cattle, and imported broken-coloured, large, horned, high-milking cows developed in the Low Countries and the Italian Po valley. There is also evidence that, at this time, some large landlords in the midlands and East Anglia tried to breed improved stock, and to practise a kind of commercial cattle ranching, with the aim of supplying working oxen for arable areas in the north midlands and beef to feed the English garrisons keeping the peace along the Celtic fringes. The purchases of winter rations by castles in south and central Wales were considerable; in 1300 Llanbadarn castle bought 155 cattle, Dinevor 125 and Cardigan 30. Towns were also becoming real establishments which required provisions, so movement of cattle across the country began and, with it, the birth of the droving industry and the cattle market.

The earliest markets coincided with traditional fairs or religious feast days. People came in from the surrounding country to pray and stayed to trade. Consequently trading usually occurred near the church and not infrequently in the churchyard; the church rises from the centre of the market place at Market Harborough. The most important markets took place in late summer and autumn when the harvest had been reaped, the beasts were in the best condition to travel and people were beginning to think of winter provisions. The four great mediaeval fairs were held at Stourbridge in Cambridgeshire, St Ives in Huntingdonshire, St Giles at Winchester and Smithfield in London. In time markets proliferated throughout England and Scotland and began to play a major role in the social as well as the economic life of the country. Entertainments went hand in hand with business and people came from far and wide for the amusement.

Bull baiting was a popular sport at these festivals. It was similar to bear baiting except that the bull was not usually tethered but allowed the freedom of the bull ring to defend itself against the taunts of mastiffs and bull dogs. The bull's death was not an inevitable climax and many lived to fight again, as cossetted in between fights 'as any English race-horse of today'. A circus for bull baiting was opened in 1570 beside the Beargarden at Bankside in Southwark and received aristocratic and royal patronage until, in 1835, the sport was declared illegal. The last bull bait was performed surreptitiously in 1853 in the village of West Derby, now a suburb of Liverpool.

Bull running was another festive sport said to have begun in Stamford in the 13th century when one, William de Warenne sighted two bulls fighting in one of his fields. 'Some butcher's dogs attacked them and chased one right through the town', and this so delighted de Warenne that he gave the field to the butchers on condition that they provided a bull each year for the running'. Every November 13th the shops of the town were barricaded, business stopped and the prettiest girl selected as a 'Bull Queen'. Then a mad bull was loosed to be chased by 'bullards' through the streets. Exhausted, the unfortunate animal was eventually killed, its meat roasted and distributed to the poor while the bravest 'bullard' was presented with the 'Great Gut or Pudding, commonly known as Tom Hodge'. In 1840 protests by the RSPCA succeeded in banning the sport but bulls are run to this day through the streets of Pamplona in Spain.

In the late middle ages the accounts kept by the large ecclesiastical and lay estates make increasing references to sales and buys at market. In 1286 Leicester Abbey bought three oxen and a bull from Loughborough market, ten oxen from Derby and other animals from Leicester. The markets also encouraged the small farmer, who formerly had little reason to expand his herd beyond his own requirements, to produce improved, surplus stock for sale.

Beasts were still mainly of indeterminate colour and conformation and of no particular breed as purity of blood was not a concern of the average 16th century farmer. But by the beginning of the 17th century some large Tudor landowners began to take a more particular interest in their cattle. Pastures were improved and with them the weight and size of cattle. In 1500 their average weight was no more than 320 lbs but by the beginning of the 17th century it was stated that oxen supplied to the Prince of Wales' household 'should weigh 600 lbs the four quarters'. The cattle were also improving in looks. A writer named Rathgeb talks of the 'beautiful oxen and cows' of England and comments on their large horns and black colour but says they are still not as large and heavy as Burgundy cattle. Another writer called Harrison eulogizes the English cattle which 'for greatness of bone and sweetness of flesh yield to none other' and says they were frequently as tall as a man of medium height and the tips of their horns were a yard apart. Their milking capacity had increased and the dairy season was extended from the beginning of May to the beginning of April and finished at the end of November rather than at Michaelmas. Calves were weaned at fifty days which is much the same as today. Modern calves usually stay with their mothers for four to six weeks and, to encourage the milk to flow from her udder, butt it, then suck one teat dry before moving quickly on to another until the calf is full or all four teats are dry.

During the 16th and 17th centuries stockmen became increasingly disatisfied with the old ways of management, and old types of cattle, and sought new methods and new animals. This restlessness was partly prompted by recently acquired notions that a cash profit could be made from animals and also by a change in the market. Instead of wool being the most lucrative part of the farm, meat, butter and cheese began to fetch equivalent prices. 'The cow became a breeder of meat and giver of milk and began to lose its primary importance as a begetter of beasts for traction'. Its place in the plough was more frequently taken by a horse especially on lighter soils where its greater speed could be exploited to full advantage. On heavier clay ground the ox continued to be used.

'It's by the mouth o' the cow that the milk comes' runs a canny Scots proverb and into that mouth must go the right food. In the 17th century cows were expected to live on grass all summer and mainly on straw in winter. Their milk yield reflected this casual attitude to feed. A Berkshire farmer called Loder calculated that the yield of a cow in the 129 days from Whitsun, on the 24th May, to Michaelmas, on the 29th September, was 129 gallons but was shrewd enough to realize that with better pasture it might produce 550 gallons a year. In the 20th century each cow's yield is carefully studied and they are fed accordingly. A cow's udder starts spurting milk the moment her calf is born. First comes colostrum, a thick, yellow nutritious

91

substance formerly called the Firstings or Beastings. After four to six days this ceases and milk begins to flow, the amount steadily increasing for three to eight weeks. During this period she should be fed more than is necessary for her daily maintenance and production to enable her to reach her peak and hold it for six to eight weeks. After this time her yield drops daily until it dries up completely. With the drop in milk goes a drop in her feeding requirements and she is correspondingly given less. For the six weeks before the birth of another calf she stays 'dry' and a good farmer will 'steam up' the cow with rich feeding to build up the developing foetus and the cow's body reserves. This will also help her milk production during the next lactation. To keep his milk supply flowing the farmer staggers the birth of calves and feeds concentrates, hay, silage and roots to supplement the cow's main diet of grass.

Cows eat an average of 70 kg of grass each day, preferring sweet-tasting plants and avoiding unpleasant-looking, hairy ones. It is almost impossible for cows to graze short grass as their method of eating is to wrap their tongues round a tuft and draw it into their mouth and with their teeth tear it from the roots. The grass is hardly chewed before being packed away for storage in the rumen. After about an hour's grazing, the cow stops for about an hour to regurgitate, masticate and swallow the accumulated food before it goes for processing through her four stomachs. Cows graze for up to nine hours in every twenty-four with peak periods before sunrise, at mid-morning, in early afternoon and at dusk and will cover about four kilometres in search of the most delectable grass. Cattle also consume vast quantities of water, between 14 kg and 25 kg a day, but since milk is 88% water this is hardly surprising.

The quality of the grass determines the quality of the milk. In spring grass gives the most milk, when it is leafy and rich in protein, sugars and water with little fibre. Every grass, moreover, has its own flavour and composition and these influence the flavour and composition of butter and cheese. Also from the air come the cultures that determine the type of cheese, which is the reason certain cheeses were only made at certain times of year. Dorothy Hartley says that

> moorland cheese was only made in spring, when the winter, hay-fed cattle went out and grazed the scented alder and new grass on the mountains; this diet cleared their blood, made delicious milk (full of casein and flavours that make for the best cheeses), and the cheese made from this special milk was 'moorland' – it was blue moulded and light in texture, and had a delicious aroma. The moorland farms never bothered to make cheese *except while the cows were on this fresh grass*.

Cheeses made on this kind of premise grew in variety through the centuries. The blue-veined, buttery Stilton became synonymous with the midlands and the Bell Inn at Stilton on the Great North Road. A softer, blue cheese was made at Cottenham and another, 'white as chalk with a royal blue vein' called Blue Vinney came from the skimmed milk of Dorset. Cheddar was a result of the chalk grazing of the Somerset Mendips and Caerphilly, a whole-milk cheese sold at ten to eleven days old in a 'green state', gained its character in Wales. The Wensleydale valley produced two, a white one made in spring and autumn and a blue one manufactured in summer between June and September. Gloucester gave its name to a small white cheese made during spring and summer and to another red, pungent-tasting one double the size. The country north of the river Ribble produced the best Lancashire cheese and the lowland country of Ayrshire the best of a similar one called the Dunlop. Fat women were said to be much in demand as wives by the farmers of Ayrshire for the weight they could put on the cheese vats. The softest cheeses came from the soft regions of Devonshire and Cornwall and the hardest from the stringent climate of Suffolk.

In the 17th century each part of the country also produced its own butter and this was sold in round pats formed by wooden moulds. Dorothy Hartley in her inimical way tells how

> Each portion was stamped with the pattern of the farms where the butter was made. Some of these old butter stamps were very finely carved, and the design, as a rule, showed the type of farm. Thus the 'swan' means a farm with water meadows; sprays of bog myrtle, a hill farm; corn sheafs, a corn-growing farm; or, sometimes a prize cow or bull set its portrait on the produce of that dairy.

Because low-lying, lush, growthy pasture gives more milk dairy cows became concentrated in certain parts of the country such as Somerset, Gloucestershire, Lincolnshire and Cheshire. Devon had its own red breed whose creamy milk made 'Clouted Cream'. The flat, sodden pastures of Essex produced a milk made mostly into cheese while in neighbouring Suffolk, Reyce said, 'large dairies . . . especially in the East . . . doe here thrive' composed of ungainly, dun-coloured cows whose milk was sent to London.

Before the railways facilitated the transportation of milk the surplus on most farms was made into butter and cheese. It was only in areas round London and other large towns that it was possible to keep cows for liquid milk production. A Mr Harrard who lived at Baumes near Hoxton kept 'three, sometimes four hundred' cows for the milk supply of Londoners and for 'victualling ships'. In the Fens 'above 2,000 milch cows, besides a great running stock' or followers were grazed on the common land of the villages of Cottenham, Chatteris, March, Wimblington and Manea and the grazings of Kent also supported large dairy herds for the London market. It was in the areas of Lincoln and Kent that the first major improvements to dairy cattle were made by the introduction of 'pied cattle' from Holland. The infiltration of Dutch cows into herds began in the 16th century but by the late 17th and early 18th it was having a substantial effect on the local cattle. In the 16th century Markham described Lincolnshire cattle as being 'for the most part Pide, with more white than the other colours, their horns little and crooked, of bodies exceeding tall, long and large, lean and thin thighed,

93

strong hooved, not apt to surbait'. But by 1707 they had completely changed as Mortimer praises them as being, 'the best sort of cows for the pail, only they are tender and need very good keeping, are the long-legged, short-horned cow of the Dutch breed which is to be had in some parts of Lincolnshire, but mostly in Kent'. From these places the cattle spread west. The Dutch cow was larger and milkier but less hardy, required more food and a far higher level of stockmanship. These factors further helped dig English methods of cattle management out of their mediaeval rut.

In other parts of the country where the land was mountainous herds of beef cattle emerged. Compared to modern beef cows, who are ready at fourteen to eighteen months of age, the Tudor bullock fattened incredibly slowly. Heifers made 'very pretty beef' at three years but steers were four or five years old before they were ready. The skeletons of young beasts grow fast, muscleing out as the animal ages and eventually laying on fat. But some breeds which grow fast mature late, while others equally fast-growing mature early. On hill and marginal land where the weather is cold, wet and windy and the grazing poor, slower-growing breeds such as the Galloway, Highland and Welsh Black are best suited to cope with the poor conditions. Their food intake is low and their full growth is not achieved until the animal is put through a prolonged store period before being finished at two years or older on food high in nutrients. On semi-intensive and intensive systems, where much rich food is available, rapid growing but late-maturing breeds such as the Hereford, Aberdeen Angus, Sussex, Lincoln and Charolais are chosen. Early-maturing beasts would finish at eighteen months or younger but be over-fat and light in weight. Few cows today end their days on the farm of their birth. Hill cows are

> bred on poor hill pasture, which gives them a strong constitution and firm bone, so that they will be good feeders (good 'doers', they call it) and put on flesh rapidly when brought down to lower, richer pastures. Bringing these young bullocks down and getting them into good condition is a separate department of farming. (Dorothy Hartley – *Food in England*)

A lack of knowledge made everything simpler for the 16th century husbandman. All he knew was that the black cattle of Yorkshire, Lancashire and Staffordshire with their large white horns tippe were the best cattle 'for meat'. That the red cattle living in Nor the moors of Dartmoor and Exmoor gave better meat than th counterparts. That even the little black runts from Cornwall moors and suffered to 'run wild in their woods and waste gro surprisingly large when taken east to Dorset or Somerset and g of their pastures. And that in Hereford, a cross between a red long- cow and some Dutch stock had produced a good type of beef cow.

Further west, in Wales, the indigenous Celtic black cattle that had pushed into the hills along with their masters by successive waves of Rom Angles, Saxons and Jutes, began to move east again in large numbers und the impulse of the meat market. The old system of spending the summers on the hills and the winter in the valleys had been eroded by the encroachment of the English and their farming habits, the disruptions caused by the revolt of Owen Glendower between 1400–15, and the depopulation caused by the Black Death. Holdings were consolidated in either valley or hill. But those in the hills found themselves deprived of places in which to winter their stock and started driving them to one of the new markets in autumn to sell to lowland farmers. 13th century records exist indicating that farmers from 'the parts of Wales' were taking their stock to Gloucester market. By the 15th century the trade had become a flourishing business. Henry V ordered as many cattle as possible to be driven from Wales to the Cinque Ports for fattening to feed his armies fighting in France. By the 17th century Welsh cattle were making their way further and further east. In 1624 the Toke family of Kent brought 30 Welsh heifers and steers at Maidstone market while in 1670 a Welsh family called Wynn sold 6 runts and 32 cows in London, 8 runts, 41 cows, 12 heifers and 76 calves at Bush Fair in Essex for a total of £297.4s.6d. and the following February 219 beasts went at Uxbridge and Maidstone markets for £470.13s. The trade became vital to the Welsh economy and during the Civil War gentlemen in North Wales petitioned the government to allow a safe passage for their cattle through England complaining that 'many thousand families in the mountainous part of this country who sowing little or noe corne at all, trust merely to the sale of their cattle, wooll and Welch cottons for the provision of bread'.

The droving trade reached its full strength in the 18th and 19th centuries. In the 17th the tributaries of cattle that gathered at fairs in Wales, Scotland and Ireland to form a river flowing into England had yet to gain their full momentum. However the trade was already providing an incentive to improve the pastures and breeds of cattle whose numbers were rising. Gregory King estimated that in 1696 there were 4½ million cattle in England and Wales alone; in 1954 there were not more than 8 million. The criteria for most was still that they should be large, big-boned, long-backed and wall-sided because, as Lord Ernle succinctly puts it in his classic book *Farming Past and Present* (1912), 'length of leg was necessary, when animals have to traverse miry lanes and "foundrous" highways, and roam for miles in search of food. Size of bone served the ox in good stead when he had to draw a heavy plough through stiff soil.' But beef was overtaking the plough in importance and the former breeding policy employed by most farmers of a 'haphazard union of nobody's son with everybody's daughter' in which cattle took six years to mature was fast becoming outdated.

During the 18th century the population of Britain doubled to just under ten million. There was a new hungry town person pumping out the manufactured goods that made the Industrial Revolution who needed ready supplies of meat. Robert Bakewell, the recalcitrant farmer of Dishley Grange in Leicestershire, was perceptive to this market change. He had already made his name with the new Leicester sheep and his showmanship and open days attracted the publicity necessary to perpetuate his ideas. But Bakewell acted more as a catalyst for breeding experiments in cattle begun several years earlier in 1740–50 by Welby, a blacksmith from Burton-on-Trent, and Webster of Canley near Coventry. In 1745 Bakewell bought two red and yellow heifers from Webster and a bull from Westmorland and began his own experiments. The standard he set for the ideal animal was 'a beast which will weigh most in the most valuable joints, not 50 stone divided into 30 stone in coarse boiling and 20 stone in roasting pieces, but vice versa' and be small-boned because 'the smaller the bones the quicker it will fat'. By mating male and female animals who visually approached these types on a 'like-to-like' basis, which frequently meant breeding brother to sister and mother to son, and with controlled out-crossing to bring in other characteristics by 1770 Bakewell had found and fixed his ideal traits. In the meantime he was forced to endure much criticism from farmers who called his practices incestuous, sinful and unscriptural and prophesied disaster.

However, others were happy to hire the results and Arthur Young wrote in *Eastern England* that Bakewell's bull 'Twopenny leaps cows at five guineas a cow'. Still others followed his example and an epidemic of breed improvement began in which the new Longhorns, as they were called, fetched enormous prices. When the Rollright herd was dispersed in 1791 it was the first pedigree stock sale in livestock history and prices averaged £85 a head, four times higher than the average fetched by animals of other breeds and comparable merit. But in his efforts to produce beef Bakewell had neglected fecundity. The Longhorn, once noted for its milk, ceased to have enough even to suckle one calf. Before the end of the 18th century the Longhorns lost favour and beef breeders looked for another cow to perfect. The one they chose was the Shorthorn.

This cow originated in Lincolnshire and there was a strong streak of Dutch blood running through its veins which enabled it to produce up to seven gallons of milk a day. Furthermore, the Shorthorn had Scotch blood which made it extremely hardy. In 1786, according to that great breeder, George Culley, the cow was to be found 'from the Southern extremity of Lincolnshire to the borders of Scotland' and was known by such different names as the Yorkshire, Craven, Durham and Teeswater. It was in Teeswater that Charles and Robert Colling, after listening to Bakewell, began to apply his methods to the Shorthorn. Their stock bull, Favourite, bred the famous 'Durham' or 'Wonderful Ox'. In 1801, at five years of age, the Durham weighed 1 ton 7 cwt and was sold to Mr Bulmer of Harmby in Yorkshire for £140. Mr Bulmer constructed a special carriage for the ox and travelled with him for five weeks before selling both ox and carriage to Mr John Day for £250. It was a bargain, Youatt says that on the same day Day could have resold the ox for £525, on 13th June for £1,000 and on 8th July for £2,000. Mr Day resisted these offers and travelled with the ox for five years exhibiting him to the public and giving personal interviews to farmers until, in 1807, it dislocated a hip and had to be destroyed.

The fame of such enormous bulls spread, until Culley remarked in 1786 that 'Nothing would please but Elephants or Giants'. By concentrating on extreme size some Shorthorn breeders repeated the mistakes of the Longhorn breeders and neglected milk supply in the race for beef. Others saw the Shorthorn's real attributes and were judicious in their improvements so that today there are four distinct Shorthorn breeds: the Beef; the Whitebread – also a beef breed but white in colour and local to the border area of Scotland and England; the Dairy – founded by George Coates in 1822 as a dual-purpose animal producing good beef steers and milky cows; and the Northern Dairy Shorthorn – a thrifty cow capable of giving beef and milk under bleak conditions. The Northern Dairy Shorthorns are direct off-shoots of the old Teeswater and, isolated in the dales of the Pennines, have remained pure.

Although the Shorthorn and Longhorn were the cows in the limelight the ideas expounded by their breeders permanently changed the course of all cattle breeding. In other parts of the country people formulated their own breeding policies and used their local cattle as foundation stock. For the first time characteristics were fixed and cattle almost everywhere could be described as breeds, each region producing its own. The red, middle-horned cattle of the west became many breeds. Among them were the Gloucester and the Glamorgan – now extinct – both of them mahogany-coloured, sleek-coated cattle with a white 'finch' stripe down back, tail and belly. Another was the Hereford, a breed which collected its dominating white face from a Dutch Blaarkop cow and its hardy, beefy character from a breeding policy set by Benjamin Tomkins of Kings Pyon in Hereford. Left a legacy of 'one cow called Silver and her calf' Tomkins created a breed which became the beef basket of the world, adapting itself to the frozen wastes of Canada, the scrub of Australia, the Argentinian bush and the South African veld. Further south another red breed, the Devon, known as 'red ruby' from its cherry-coloured, curly coat was famous as 'the speediest working oxen in England' able to plough an acre of heavy soil a day and produce rich milk and beef. In time the breed divided into the South and North Devon. The former became a dairy cow. The North Devon developed into a beef cow and it was the one that drew 'the plough which turned the first furrow in the soil of New England'. Sussex cows were a related breed, also red and noted as plough oxen. Teams of Sussex oxen moved the heavy weald clay and hauled timber to fuel the Sussex iron-smelting industry from Saxon times, developing in the process immense muscles ill-adapted for anything else. Attempts were made to refine the breed, make them 'more kindly feeders', reduce their size and 'render the bone finer and improve the form' by crossing in Shorthorn blood and today the Sussex converts pasture from Pevensey Marsh to the South African veld into prime meat.

Around Exeter and in the Isle of Wight there were cattle imported from the Channel Islands of Alderney, Guernsey and Jersey. Small, pretty, delicate animals they yielded rich, creamy milk but were too soft for most British farmers and Culley says were generally kept 'as appendages to gentlemen's parks or rural villas'. Sheeted or belted cows, that is black and red cattle with a broad band of white round their middle, were also observed by Marshall in 'Gentlemen's seats in various parts of the island'. These sheeted cows were almost certainly derived from importations of Dutch Lakenvelder cattle. They were thought to have come over first in 1690 from Holland, brought by Sir William Temple. As white belts are a dominant colour gene, cows with 'white sheets thrown over their backs' are easily bred without disturbing any other qualities. Thus Sheeted Somersetshire, Belted Galloways and Belted Welsh all became decorative derivatives of these breeds perpetuated, except for the Somerset which is now extinct, to this day.

Outside the park gates, in the towns, the problem of how to feed their ever-expanding populations was a cause of great concern to contemporary farmers, and politicians. The consumption of meat by the average person was estimated to be half a pound a day and although breeding had improved the size of cattle from about 370 lb in 1710 to 550 lb in 1795, not enough were bred to feed the ever-open mouth of London and other large towns. At the same time the possibilities of winter fodder crops began to be fully explored. 'Turnip' Townsend enthusiastically planted this root in Suffolk while others experimented with carrots, potatoes, cabbage and oil cake – an idea borrowed from the Dutch who reprocessed rape seed, once the oil had been expressed, into concentrated cake. In the 18th century the practice of droving cattle from Wales and Scotland to fatten on these crops before being sold in London became established and reached a peak.

In 1778 Pennant reckoned that 6000 cattle swam the Menai Straits from Anglesey, 3000 started from Lleyn while many others crossed the border from Pembroke, Carmarthen and Glamorgan on their way to London. The droves followed cross-country routes, often using ancient trackways.

Irish cattle were also included in western droves. These 'Whicke Kayne' or black cattle were similar to the Welsh and, according to Sir William Temple writing in 1673, had 'never been either handled or wintered at hand but bred wholly upon the mountains in summer and upon the withered long grass and the lower lands in winter'. Their import began after an attack of murrain had temporarily stopped the flow of Welsh cattle in 1632. From this time on, apart from brief stoppages because of complaints by west and north country breeders anxious to protect their interests, these black cattle continued to flood in through the west coast ports.

Cattle from Wales were in charge of drovers who could only get a licence for this responsible position if they were married men, householders and over thirty years of age. It took some skill to negotiate a herd, usually numbering about 200 beasts, across the mountains and rivers of Wales, past the robbers and highwaymen of the borders and bring them safely to Smithfield on the day before the weekly market. The drover received the cattle 'in trust', and paid for them on his return but the danger of being beset by highwaymen and robbers was great for men carrying large sums of money and the consequences dire for the farmer dependent on the cash.

The Black Ox bank, now part of Lloyds, owed its beginning to people attempting to find a better system for dealing with the cash.

After the Union between England and Scotland cattle began to pour across the border from the north. The cattle were probably a mixture of types mostly made up of 'black homyl' or polled cattle similar to the Galloway or Aberdeen Angus, a few of the remaining 'wild milk-white Bulls', and the shaggy, red precursors of the Highland cow, the Kyloes of Argyll and Wester Ross, so wild they 'flees the cumpanie or syght of man' but whose flesh had 'a marvellous sweetness and tenderness'. Added to these were the small, black, succulent-fleshed cattle of the Western Isles, probably akin to the Welsh and Irish breeds. In 1663 the cattle passing through Carlisle alone amounted to 18,574. They undercut the north country beasts though swingeing customs duties were imposed of £10 a head for cattle and £5 for calves.

More than half the Scots cattle assembled at the Falkirk Tryst. They made their way there from all over the mainland and islands; those from Skye began their journey by swimming across the Kyle. The Tryst was held three times a year in August, September and October. Thomas Pennant in 1772 described the town as a 'large ill-built town supported by the great fairs of black cattle from the Highlands, it being computed that 24,000 head are annually sold'. From here the cattle began their slow trek south moving by many routes across the green hills of the borders. If they left the soft, grassy tracks for stony land or the hard highway they were fitted with 'shoes' similar to those worn in Roman times. Once across the border some of the droves went south down the east coast through Durham while others travelled a western route via Carlisle; their numbers being swelled by cattle from Galloway and Ireland and later by those from Wales. The paths which followed these two main routes were fragmented into a complex network and were often changed because of political or social conditions, market requirements and the drovers' own preferences. Droving tracks can usually be recognized by a number of parallel paths (sometimes covering a width of twenty to thirty yards) made either by beasts going in straight lines or by successive droves moving to new ground to avoid that churned up by their predecessors. These ridgeways or green roads cross the Hambleton Hills in Yorkshire, the Cotswolds, Chilterns and Mendips and many are of ancient origin.

The droving traffic increased until, during the last part of the 18th century, something like 100,000 cattle and 750,000 sheep, besides pigs, poultry and assorted pack horses trekked to Smithfield alone and the unmetalled highways became a 'perpetual slough of mud'. As a way of paying for paving, the first Turnpike Act was passed in 1663, and between 1760 and 1774 as many as 452 were put through Parliament to the benefit of the general public but the ire of the drovers. The metalled roads they paid for at between tenpence and eightpence 'for every score of oxen and neat cattle' have been an everlasting asset to Britain.

Scottish drovers were often shaggy, wild-looking men who slept rough with their cattle. They wore coarse, homespun tartans and 'Scotch caps', carried their own basic rations, mostly consisting of a few handfuls of oatmeal with which they made a kind of porridge called 'crowdie', one or two onions and a ram's horn filled with whisky. The droves were accompanied by 'Coally dogs' of many shapes and sizes. After the journey south, if the drover travelled home by carriage or was detained by business, the dogs were turned loose to find their own way back; following the same route in reverse and being fed at inns or farms where the drove had stopped. The drover paid for the dog's food on his next trip.

Songs to please men and beasts played an important part in droving lives. Some songs were devotional hymns or commemorated St Brigit, the patron saint of cattle, dairy maids, fugitives, Irish nuns, midwives and newborn babies, St Brendan, the patron saint of voyages, or St Columba who looked after the cattle of the Western Isles. Other ballads passed on the news of untrustworthy drovers, noteworthy fairs or sweethearts left behind, such as this one:

> I lost a stirk on the shores of Loch Linnhe,
> Two stots at Salen Loch Suinart,
> O'er by Loch Shiel I've lost my heart
> To Jean, a lassie of Moidart.

Scottish drovers operated under a different system to the Welsh. They generally paid for their cattle before setting off, or got credit, and took the risks involved. The average price for a small Highland cow at the end of the 17th century was £4 so the outlay was considerable. Profits per head of cow seldom amounted to more than two shillings and sixpence to five shillings, and this profit could be decimated by disease. It happened to Benjamin Bell and his son Thomas who set out from Canonbie in Dumfries for East Anglia in 1745. Writing home to the provost of Annan, who may have financed the drove, Bell laments that:

> The distemper amongst the cattle rages more and more; it is now over almost all Norfolk, Suffolk and Essex, so that we have no place to fly to, even if we had liberty. God knows what we shall do. We cannot get money to bear pocket expenses; all manner of sale is over; our beasts drop in numbers every day and we have an express from William Johnston in Essex that the distemper is also got among our beasts there and dying in half dozen and half scores every day; our conditions are such that several drovers have run from their beasts and left them dying in the lanes and high ways and nobody to own them . . . there is upwards of 300 lost in one hand here already . . . I cannot half express our melancholy situation. May God pity us.

In this next letter Bell draws a sad end to his story;

> All is over now. We can neither pay London bills nor nothing else. We

100

have about £1000 of charges to pay in this country and in the road and not a shilling to pay it with.

During their journey to London cattle could frequently change hands four or five times. Trow-Smith tells that in extreme cases a Welsh bullock could

> be bought at a Caernarvonshire market from a hill breeder as a yearling, be run on to three or four years of age by a valley grazier who would then dispose of it at market or privately to a drover who would walk it into the English Midlands. Here it could be sold at, perhaps Northampton market to a Harborough grazier who would nearly finish it and dispose of it to a Home Counties turnip man around Hatfield, who would fatten it and sell it finally at Smithfield to a London butcher. (*Livestock Husbandry*)

Others tramped straight to East Anglia and the great gathering fair at St Friths outside Norwich. It was held in October and lasted several weeks or until there were no more cattle to sell. These cattle were mostly passed on to graziers who presented them with pasture way beyond their previous experience. The Rev William Gilpin commented that

> Of such pasturage they had no idea. Here they lick up the grass by mouthfuls; the only contention is which of them can eat the most and grow fat the soonest. When they have gotten smooth coats and swagging sides they continue their journey to the capital and present themselves in Smithfield where they find many admirers.

A few ended their journey at Barnet Fair but most made their way through Highgate, Islington, the Holloway Road and Upper Street to Smithfield in droves 'half a mile long'. Cattle from the west progressed down the Edgeware Road, Oxford Street and the Marylebone Road.

Smithfield market, 'without question' says Defoe 'the greatest in the world' processed, at its height, a weekly average of 69,946 sheep and cattle. The market took place on Monday and Friday amid a welter of noise and blood. Dickens conjures up the scene in *Oliver Twist*:

> The whistling of drovers, the barking of dogs, the bellowing and plunging of oxen, the bleating of sheep, the grunting and squeaking of pigs, the cries of hawkers, the shouts, oaths and quarrelling on all sides; the ringing of bells and the roar of voices, that issued from every public house; the crowding, pushing, driving, beating, whooping and yelling; the hideous and discordant din that resounded from every corner of the market; and the unwashed, unshaven, squalid, and dirty figures constantly running to and fro and bursting in and out of the throng, rendered it a stunning and bewildering scene, which quite confused the senses.

In 1858 'the putrid blood running down the streets, and the bowels cast into

101

the Thames' caused the population to complain, the live market was moved to Copenhagen Fields in Islington and Smithfield became the dead-meat market it is today.

Droving was mainly killed by the railway as this could transport animals quicker and cheaper. Other factors which contributed to its demise included the enclosure of land, which made it difficult to follow old tracks and take advantage of ancient rights to free stances of grazing – and high turnpike charges. The English graziers also started to complain about the high prices asked by drovers for beasts of poor quality which had lost weight with every yard of their journey. This led them to start breeding their own stock. Large-scale dairy farming also got under way at the beginning of the 19th century and became more profitable than fattening cattle for beef. By the start of the 20th century droving had ended.

The railway also killed the urban cow. Until they facilitated the fast transportation of liquid milk in the early 19th century all large towns had their dairies. Liverpool, for instance, kept between 500–600 cows in her centre but London had far the most, its population was estimated to be 8,000 of which 3,950 were housed in the area bordered by Paddington and the Grays Inn Road, about 750 in the West End and the rest in the suburbs and East End. Town cows were fed on brewers' waste and the produce of suburban farms which existed to grow their food. Shorthorns were the favoured breed and the cows were kept for one lactation only before being sold on. They were tied in twos in cowhouses and let out into yards for exercise.

The enclosure of common land, that accelerated throughout the 18th century, destroyed the cottage cow. Agricultural improvers were affronted by 'the very wretched appearance of the animals who have no other dependence but upon the pasture of these commons who, in the most instances bear a greater resemblance to living skeletons than anything else', and opined that five acres of a peasant's own land was of far greater benefit than 250 acres of common land, frequently over-run and over-grazed by other villagers' animals many of which were diseased. Communal grazing gave no control over the breeding of animals and consequently there was no incentive to correct their form. However, deprived of his common rights the cottager was also deprived of a certain independence. Forced to buy what he had formerly grown for himself but without the means to earn the money he took to the bottle, to the life of a robber or highwayman or drifted into the towns.

Enclosures were not all bad. There is no doubt something had to be done to remedy the worst situations where cattle were not only starved and diseased but housed 'huddled together into draughty, rickety sheds, erected without plan, ranged round a yard where the liquid manure, freely diluted from the unspouted roofs, ran first into a horse pond, and thence escaped into the nearest ditch'. Enclosures also encouraged the drainage of land. It is difficult from the dry standpoint of the 20th century to appreciate quite how sodden England was before drainage. Huge areas of Somerset, Kent and East Anglia were almost permanently submerged. A reporter of 1794 mentions that in the Lincolnshire fens 'the occupiers frequently, in one season, lose four fifths of their stock' from disease and drowning, but once drained 'the benefit resulting therefore is astonishing and thousands of acres of the Brent Marsh in Somerset 'heretofore overflown . . . and of little or no value are become fine grazing and dairy lands'.

Numbers of cattle rose from under five million in 1866 to over six million in 1874 and reached the 'highest standards of excellence in symmetry and quality'. Theories of heredity and of dominant and recessive characteristics discovered by the Austrian biologist and monk, Gregor Johann Mendel, helped correct some of the worst abuses of continuous in-and-in breeding, enabled farmers to fix new variations, blend useful characteristics in a single type and establish new strains. Colour marking was one of the first contributions of Mendelian genetics to cattle breeding. Certain breeds pass their colour on to their offspring whatever breed they are crossed with and thus the paternity of their progeny is obvious. The white face of the Hereford and the black coat of the Aberdeen Angus are always transmitted whereas the Shorthorn leaves no such identifying mark.

All these theories, experiments and different public demands led cattle to become specialized into three types, beef, dairy and dual-purpose. The general utility animal where 'the bullocks worked and beefed, and the females milked, worked and beefed' gradually disappeared. Today the dual-purpose cow is also a rarity. The dual-purpose cow is epitomized by the old Shorthorn, but a close relation, the Blue-Albion, a blue roan-and-white cow found in the north midlands, and two Irish breeds, the Dexter and Kerry both imported to England in the 19th century were also popular. The Dexter is a miniature-sized black or red cow developed by smallholders in Ireland as house cows subsisting on the minimum of feed and giving the maximum of milk and beef. The Kerry, rather 'higher in the leg' and more graceful in the body and horn, is a relic of the old Celtic black breeds and will exist on so little food it is said that if you put green spectacles on them they will feed off the main road. The latter three breeds all fell into decline, but have recently been resuscitated by the Rare Breeds Survival Trust and are beginning to find their feet again.

The beef breeds established in the 18th century, the Hereford, Sussex, North Devon and Galloway, all reached new heights of perfection and prominence in the 19th and a newcomer joined their ranks from Scotland: the Aberdeen Angus. This polled, glossy black beast from the north-east began to usurp the position formerly held by the horned Banff, Buchan, Mearns, Falkland and Fife cattle. The Aberdeen Angus filled its compact form with excellent meat on the barley and rich grazings of the north-east and it became the most rapid beefing animal in Scotland. These attributes have taken it to most of the beef producing countries of the world.

On the west coat of Scotland another beef breed established itself during the 19th century. The Highland cow evolved from a melting pot of types on the west coast and became fixed as two kinds. One, the more primitive, was found in the islands and known as Kyloes. These were small, shaggy and mainly black while the cattle on the mainland were equally hairy but variously yellow, red, brindled and black and with horns like buffaloes. The improvement of the Highland cow began in the 19th century when certain landlords such as the Dukes of Argyll and Hamilton, the McNeils of Barra and the Stewarts of Harris became anxious to improve the prices their cattle fetched in London. They made the Highland what it is today, a breed noted for its longevity, hardiness and subsistence on scant pasture but slow to mature and it is now frequently crossed with the faster growing Beef Shorthorn to make a new breed, the Luing.

Cattle were intrinsic to the Highlander's way of life. They still practised the ancient system of 'transhumance' and spent the winters in the valleys and the summers in the hills, driving the cattle

> to the inland glens and moors, which are covered with hard grasses and rushes, because the portion that yields soft grass is not sufficient for their consumption during the whole year.

The summer was the happiest part of the Highlander's year as winter was spent closeted with their cattle in 'miserable huts', the traditional 'black house' where, Youatt says,

> The fire is placed in the middle of the floor. The soot accumulates on the roof, and, in rainy weather, is continually dropping, and for the purpose of obtaining it for manure, the hut is unroofed in the beginning of May. The dung of the cattle which had been accumulating during the winter and spring, and had been mixed with straw, ashes, and other matter, is at the same time removed from the outer apartment.

The death of droving meant that the numbers of cattle were no longer in demand. Rents went unpaid and the landlords sought to remedy the state of affairs. Sheep were fetching high prices and they thought to substitute them for cattle and introduce a new system of husbandry. However, since the only joyous part of the Highlander's poverty-stricken life was the time spent in his summer sheiling he revolted at the prospect of abandoning this in favour of sheep. The landlords retaliated by trying to resettle areas with new, more tractable people. Outraged, the Highlanders attacked the incomers, broke down the new fences and drove the sheep into the lochs to drown, or across the county border. Laws were enforced to suppress 'the violence of the mob' and against this unequal strength the Highlanders gave way. Sheep covered the hills, deer forests were brought under cultivation and the cattle, reduced in number but improved in quality, became, according to Youatt, 'fatter and happier'. The human population, said Youatt were

> certainly not so numerous . . . but of a different character – more intelligent, more industrious, more respectable, more useful; and the remainder have either sought employment in the south, or emigrated

to America or some of the British colonies. The value and the rent of the land is trebled – quadrupled; and the tenant can pay it, which he could not before: while, in a national point of view, the addition of food, the increased value of stock, and the unprecedented supply of the raw material for one of our most important manufactures, are circumstances of immense importance.

The landlords were not all wrong, undoubtedly something had to be done to pull the Highlanders up from their almost neolithic standards, but in many cases it could have been undertaken in a more sympathetic fashion. The pattern they created underlies the agriculture of Scotland to this day. Cattle are concentrated on the lower ground in the west, on the corn-growing land in the north-east, in the green hills of the borders and the flat land of central Scotland and Fife, while sheep take up the rest of the space.

The incentive to reform the beef cattle of Scotland may have been the draw of the London markets but the reasons for reforming the only notable dairy cow were the dense populations of Paisley, Greenock and Glasgow. These consumed all the milk, butter and cheese the dairy farmers of Ayrshire, Renfrew and Dumbarton could produce. The Ayrshire cow was based on the old Scottish breeds with additions of other blood. It probably owes its large milk yield (up to 900 gallons) to Dutch influence, its richness of milk to Channel Island blood and its thriftiness to the Longhorn. The breed was originally called the Dunlop after one, John Dunlop who in about 1760 introduced some 'English or Dutch cows to his byres at Dunlop House'. A pair of the 'new breed' were given by Mrs Dunlop to her friend the poet, Robert Burns when he set up at Ellisland Farm.

The Suffolk Dun was another great dairy breed. Its butter, according to Arthur Young, was 'justly considered the pleasantest in England' and 'The quantity of milk they yield exceeds that of any other breed I have met with in the kingdom.' This yellow, dun or red, polled cow whose angular looks caused the derision of Bakewell, who announced it would look better turned upside-down, had been praised since the 17th century for the quality and quantity of its milk. The Suffolk was probably derived from the same ancient stock that made the Norman and Channel Island cow but its produce was derived from a system of feeding. In northern parts of Suffolk, Young notes, there is not a single dairy farm 'without crops or cabbages and turnips which were raised for the cows'. The 'demand for cattle of large carcase became general' in the early 19th century and the Suffolk was crossed with red, beef cattle from Norfolk. The result was a dual-purpose cow, the Red Poll, a breed evolved principally by one, John Reeves, a tenant of Holkham Hall. The pure Suffolk Dun disappeared and is tragically now extinct. The Red Poll, for a while, reigned supreme as the prime dual-purpose cow, but as 'jacks of all trades' ceased to be in demand, they too became rare.

During the 19th century Jerseys and Guernseys expanded their territory beyond the gentleman's park. The creaminess of their milk gained them a reputation for being among the best dairy cattle and numbers increased until they exceeded all other breeds except the Shorthorn. At the Royal Show in 1909 there were 91 Herefords, 195 Jerseys and 422 Shorthorns. However, their thin skins and delicate constitutions confined them to the south of the country. Another southern breed, the South Devon flourished throughout the dairying districts of the Vale of Exeter and Honiton and along the coast from Dorset to Cornwall. Here it thrived, says Youatt

> in spite of neglect of injury. *The ground secret of breeding is to suit the breed to the soil and climate.* It is because this has not been studied, that those breeds which have been invaluable in certain districts, have proved altogether profitless, and unworthy of culture in others. The South Devons are equally profitable for the grazier, the breeder, and the butcher; but their flesh is not so delicate as that of the North Devons. They do for the consumption of the navy; but they will not suit the fastidious appetites of the inhabitants of Bath, and the metropolis.

It may have been an Alderney or South Devon or one of 'the mongrels of every description: many of them, however, are excellent cows, and such as are found scattered over Cornwall, West Devonshire, Somerset, and part of Dorset' that Tess of the d'Urbervilles milked 'Amid the oozing fatness and warm ferments of the Var Vale, at a season when the rush of juices could almost be heard below the hiss of fertilization.' Hardy gives, in the same book, a marvellously evocative account of the cows coming in for milking from the meads.

> The red and white herd nearest at hand, which had been phlegmatically waiting for the call, now trooped towards the steading in the background, their great bags of milk swinging under them as they walked. Those of them that were spotted with white reflected the sunshine in dazzling brilliancy, and the polished brass knobs on their horns glittered with something of military display. Their large-veined udders hung ponderous as sandbags, the teats sticking out like the legs of a gipsy's crock; and as each animal lingered for her turn to arrive the milk oozed forth and fell in drips to the ground.

The cattle would have trooped into the milking parlour in the same order morning and evening as cows in a herd maintain a constant position. While grazing, one or two animals lead and the rest follow in their direction. However, curiously, once the herd is on the move these leaders in the field stay in the middle allowing the mid-field grazing cows to go first.

Cows are not noted for their intelligence. They are simple animals, estimated to be only half as bright as horses, but with acute senses, especially of smell. People attribute the cow's lack of wit to the fact that when an animal is made the slave of man and no more is required of it than to pull a plough or make its way to and from the milking parlour or stand fattening in a field its instincts and intelligence languish. The ferocity of wild bulls is commemorated in ancient sagas but once domesticated they

106

are described as 'ox-like' and made synonyms for men with more muscle than brain. 'Bovine' is applied to large, dull women, and the domestic cow is the symbol of placidness, docility and motherliness. As a meditative image the picture of a cow munching in a meadow calms the mind, while the flashy, unpredictability of a horse or pig disturbs. As Michael Denny wrote of his cow, Buttercup, in the *Observer* of May 27th, 1979

> besides producing milk cows have beneficial psychological effects on people. Cows can be a cure for neurosis – for angst, anomie, alienation, the blues, cafard, culture shock, future shock, identity crisis, the mean reds or whatever you've got. Everyone with a high-powered job should have one, preferably in the office, to keep them calm in moments of stress.

Unbeknownst to the owners of these passive beasts their whole world was about to be disrupted by their own unwitting deeds. The cattle farmers had exported to the new British colonies and the knowledge of crop growth began to return in other forms. Initially it came in the form of cheap feeding stuffs which the British farmer welcomed as they enabled him to stall-feed his cattle, raise his milk yield and make greater profits. But then the colonists started sending in cheap wheat, barley and oats and fat cattle and, after 1877, America, Canada, New Zealand and Argentina joined the trade. Meanwhile the outgoings of the British farmer mounted and, as expenses of production rose, profits fell. These were further decimated by disease brought in by the imported cattle. Rinderpest, pleuro-pneumonia and foot-and-mouth all became endemic during the 19th century destroying thousands of cattle. 10,056 were slaughtered in 1866 and in 1879 even hares, rabbits and deer died of the rot. The scourge continued until the Contagious Diseases of Animals Act was passed in 1896 which required all animals to be slaughtered at their port of origin. Veterinary skills also made vast advances; anthrax, redwater fever, foot-and-mouth and tuberculosis were all defined and steps taken towards their eradication. Some methods were drastic, such as the compulsory slaughter, still in effect today, of animals affected by foot-and-mouth and TB.

Though these successive blows sank the British farmer into a slough, by the end of Queen Victoria's reign the unassailable commodities of fresh meat and liquid milk became the answer and farmers began to streamline their enterprises. Cattle numbers were increased to supplant unprofitable corn. Housing was improved, helped by Liebig's discovery that warm cattle required less food, and additional feed was supplied by ensilage (or green grass preserved by the exclusion of air and the adding of molasses) while chaff cutters and turnip slicers made root crops easily assimilated. Milking machines came in and were universal on all farms with more than about ten cows. Cheese makers discovered the exact chemicals, bacteria and acidity that make cheese enabling them to produce it in larger, more uniform quantities. Cream separators and mechanically operated butter churns relieved much of the dairymaid's labour.

Until the First World War the milk industry expanded slowly, its rise coinciding with the growth of the urban population. Then the invention of the motor engine, which meant that milk no longer need be supplied by farms near cities but could be brought by lorry from a distance, altered and increased demand. Large dairies sprang up offering annual contracts for milk in churns collected direct from the farm. Then came the War and after it a deep world depression. Corn and sheep were losing money and too many farmers looking for security went into milk; this, coupled with the dumping of cheap Australian butter and cheese, collapsed the market. The government took a hand, instigating the Milk Marketing Board in 1933 to act as a central body setting and controlling prices and production. During its first year it sold 856 million gallons of milk, by 1946 this had risen to 1,244 million gallons, by 1960 to 9,676 million gallons and in 1980 was 15,160.

Clean milk became a concern of the consumer. Cow-sheds changed into cow-parlours, spotless, chromium-plated and air-conditioned. Robert Stenhouse persuaded farmers of the benefit of grooming, washing and sterilizing cows until they became 'better housed than a lot of Londoners and probably cleaner too'. Cattle diseases which can be transmitted through milk were seriously investigated. Lord Rayleigh was the first to sell milk from Tuberculin-tested cows and in 1914 Robert Hobbs of Kelmscott opened his shop selling Certified Milk.

Beef farmers, realizing that the British were prepared to pay for succulence, abandoned the unequal struggle to grow corn, laid their land down to pasture, and utilized the cheap foreign feed stuffs as extra fattening material for their cattle. The practice of moving animals through various farms and stages before being slaughtered was perfected. Within breeds, strains which exactly fitted these feeding systems were isolated and promoted. The newly established Meat and Livestock Commission introduced Performance Testing whereby the heaviest cattle, which are also above the average standard for the breed, are selected to pass on their genetic potential to their progeny.

This is mostly done by artificial insemination. AI, a form of 'induced masturbation', was invented by the Russians and given its first demonstration in Britain at Cambridge in 1934. The bull is first stimulated by being shown a cow in heat, then his erect phallus is manually guided into an artificial vagina made of thick, lubricated rubber surrounded by a warmed, pressurized water jacket. The sperm, collected in a sterile glass phial at the end of the tube, is carefully treated and preserved in deep frozen liquid nitrogen. Since AI means it is possible for one bull to sire thousands of calves it became vital that his sperm and subsequent offspring be of superlative quality. Sire Performance Tests and Progeny Testing began. The scientist superseded the breeder and his herd books. Today 75% of the cattle in England and Wales alone are born by AI. This equals an average of 10 cows a minute or 2 million cattle a year.

Embryo transplant is the newest instrument for the multiplication of cattle. Pedigree cows of proven ability are given super-ovulating drugs then artificially inseminated. The fertilized eggs are recovered and implanted into poor quality, proxy mothers. In this way as many as forty eggs can be taken from a cow and, in 1981, a record 19 calves were born through embryos transplanted from one Limousin cow. Some cows take unkindly to this manhandling of their systems and reject the embryo. Even so, the belief that embryo transfer will be used increasingly to spread the influence of top-quality cows was confirmed when, in 1979, £4,500 was paid for an unborn embryo, and in 1981 a syndicate gave 32,000 guineas for a Friesian cow, Ullswater Beatexus, expressly to provide embryos.

The Friesian is the dairy cow of the 20th century. Big, boring and reliable they originated in Germany. During the 18th century flooding, and cattle plague, decimated the black and white cattle population of the Zuider Zee in Holland and they restocked with cattle from Jutland. The efficient breeding methods of the Dutch made the Friesian cattle famous for their huge milk yields and, by the end of the 19th century, they were being exported to Britain. The Second World War and a great bull called Terling Marthus established the Friesian's position as the leading milk cow in Britain. During the war the government paid for milk by bulk and not by its content of solids and butter-fat. As the Friesian can produce 5441 kg of milk a year, albeit of thin, watery quality, as opposed to the Ayrshire's 4863 and the Jersey's 3776 its place was assured and gradually it has supplanted all the other dairy breeds regardless of how rich or creamy their milk might be.

Another boost was given to the Friesian's popularity by the fact that the steers and discarded cows fattened into better carcases than other dairy breeds and its bull calves made good veal. Ever since the rearing of veal ceased to be merely a way of using excess milk from the house to fatten a calf for special occasions, the production of veal has been a controversial issue and basically cruel. The industry was under way by the middle of the 17th century, when the surplus stock of the Home Counties dairymen was being fattened for veal. The 'exceedingly white and fat' meat was obtained by force-feeding the calves milk and balls of concentrated fine cereal meal or split peas mixed with milk and a little brandy, gin or aniseed water. These were crammed down the calf's throat with the aid of a stick and sometimes draughts of black pepper and urine were administered to encourage a thirst and further suckling. White veal fetched a higher price than pink veal and to blanch the meat the calves were sometimes fed chalk, bled at fortnightly intervals from the age of six weeks and killed by a barbarous process; 'they bleed him at evening, allow him nearly to bleed to death, stop the blood, and slaughter him next morning to produce the whitest flesh' told a writer called Kalm. However the calves were at least kept in communal straw yards and got real milk. It is for their housing and feeding methods that veal producers today are criticized. Calves are kept in darkened sheds, confined without bedding in crates so narrow they are unable to turn round or even lie comfortably. Since the 1950s they have been fed a milk substitute to make then anaemic, packed with drugs to prevent disease and growth-inducing oestrogens which remain in the meat and can be transmitted to humans causing bizarre side effects such as obesity, abnormal sexual development in children and cancer in adults. Deprived of roughage in their diet the calves lick themselves and hair balls accumulate in their stomachs. Public outrage at the treatment of veal calves is beginning to have an effect. Seeing sales slump producers are looking at alternative, kinder methods and experiment with keeping calves in straw yards. But with so much capital tied up in the crate system the opinion of most veal farmers was voiced by a French producer who said that until 'the EEC decides to change the law on pen size, we will resist as long as possible because we all have the pens and it would be very expensive to scrap them'.

The recent drop in liquid milk sales could affect the supply of veal calves. Milk has stopped being the most important product of 52% of farms over 20 acres. In 1979 schools ceased providing free milk removing a market for 18,260,000 gallons a year and in the same year the Milk Marketing Board switched to bulk collections only, driving 2000 small dairy farmers out of the business.

Farmers are now being offered a quality payment scheme aimed at improving the butter fat content of milk, as higher fat levels mean more butter and cheese for the same amount of milk. This attitude may well affect the breeds of cows that are kept. The situation where 76% of all cows are Friesians, 10% Ayrshires, 9% Channel Island breeds and 2% Dairy Shorthorn could be reversed.

Perhaps the richer-milked Jersey, Ayrshire and Gloucester cows will come to the fore. There is also a wholesome move back to production on the farm. Whereas the Milk Marketing Board finds that foreign competition for its mass-produced butter and cheese gets annually hotter and the home market for its frequently bland-tasting productions decreases, farmers who have opted to process their own milk into yogurt and cheese cannot keep up with demand. It may end with dairy farming divided into two. On the one hand there will be the dairy specialist with a huge herd milked three times a day, to get the maximum from each cow, using machinery controlled by a computer which enables one man to handle hundreds of cows; while on the other hand there will be the small farmer, machine-milking his own cows by himself, churning the liquid into butter and cheese for sale at his farm shop at the gate.

Beef too seems to be at a divide. Many cows still pass their lives on several farms but, unlike dairy cattle, are rarely the major enterprise of that farm, except in Scotland. Many cows kept for beef are not even beef breeds but are often a by-product of dairy farming. The younger female dairy cattle, not yet required as herd replacements, are put to a beef bull and their offspring fattened. The unwanted bull calves from the dairy herd may also be fattened. So beef cattle can also be dairy, dual-purpose breeds or crosses of beef bulls and dairy cows. The numbers of beef cattle rose by 73% between 1954 and 1970, partly due to government subsidies and the development of intensive methods and partly to the breeding of more efficient stock. In lowland Britain the out-wintered bullock has nearly disappeared, replaced by 'barley-beef' – housed for the whole of its short life of about a year – and '18-month-old-beef' which has a summer at grass sandwiched between two indoor winters. These indoor cattle are fed concentrates and barley, stored wet and preserved in its own gasses in sealed silo towers.

The quest for the ideal butcher's beast is an elusive business. Farmers experiment with different combinations of crosses in an effort to attract the judge's approval in the show ring and the butcher's in the market. A Limousin × South Devon, a Charolais × Welsh Black heifer, a Charolais × Hereford and a Charolais × Aberdeen Angus all won prizes at Fat Stock shows between 1979 and 1981. These calves acquired fast growth and leanness from their Continental parent and hardiness from their British one.

It is the Continental breeds that set the modern records for beef production. These cows are all huge, of ancient origin, and until recently were drawing ploughs in their native countries. Compared to British breeds they are more muscular and produce leaner, if blander meat but require much more food to fill their vast frames and, unable to stand cold, need housing. It was the development of intensive methods of farming and feeding that brought them into prominence. The Charolais was first imported in 1962 and, while the cows only produce enough milk to feed a single calf, their bodies achieve bulk at a fast rate. The Chianina from Italy is the largest cow of all. In 1980 one weighed almost 1½ tons – ¼ ton more than the Durham ox weighed in the 18th century. The abnormally thick muscleing of the Chianina's hindquarters may cause difficulties while calving a calf but not when it comes to carving into steaks. The brick-coloured Simental from Switzerland and West Germany, the Blonde d'Aquitaine, the Belgian Blue and the caramel-coloured Limousin are all recent imports. At the moment the French Limousin is the most popular; 'That bull is the nearest thing to a Landrace pig for leanness that we have yet seen' said someone at a sale in 1981 when the top price of 7500 guineas was paid for a bull. All the highest prices are now being paid for these 'exotics' as they are known. As John Stewart Collis cynically says in *The Worm Forgives the Plough* (1973)

> The truculcnt motorist who hoots his way through cows on the road would change his tune if he knew their money measure . . . how his jaundiced eye would moisten with greed if he found a thousand-pound cow blocking the path of his car ((itself likely to fetch thirty pounds). Probably he would raise his hat to it).

Unfortunately many of the unscrupulous have realized their value, cattle rustling is on the increase and the padlocked field becoming the norm.

'Today's meat market is all about shape,' said a farmer recently. This kind of opinion coupled with the competition from the rangy continental breeds is having a worrying influence on the shape of our own indigenous cattle. In an effort to compete, breeders, according to David Sinclair a top breeder of Aberdeen Angus, have 'swung too much to the taller and narrower type of animal'. He believes there is 'still a strong demand for beef at the top end of the market' and that the Angus is the ideal sire to produce quality beef. 'Don't turn the Angus into a black Charolais,' said another breeder. 'Some breeds like the Charolais and Simental will certainly give big live-weight gains but they need expensive energy feed to do it. The higher cereal prices become, the more important it is to look at the valuable asset we have in our breed.'

Foreigners certainly do not underestimate British breeds. In the past the main purpose of British breeders was to produce cattle for the home market, now much pedigree stock goes abroad to Egypt, Iraq, Libya and other Middle East countries. Worried about feeding their oil-rich populations they are seeking to improve their own stock and farming with thrifty British breeds. Aberdeen Angus and South Devons have recently been sold to South Africa and Australia and the present of a Galloway bull to Chairman Hua on his visit in November 1979 could ensure that this hardy animal finds a home in China. In this country people are eating less meat but beginning to demand better quality. 'There is a swing back to Angus beef which is being dictated by the housewives,' said a farmer after the smallest but best parade of Aberdeen Angus cattle in the famous Perth Bull Show's history.

For whom is all the meat produced? EEC beef mountains fill the news, their production a creation of a protective pricing policy because

> Fattening cattle in the EEC countries is far more expensive than it is in countries such as Argentina, Brazil, Uruguay, New Zealand and Australia, where there is ample space for grazing. Unlike milk production, beef production is not labour intensive but space intensive. In Europe land is dear, and animals have to be kept in byres during winter and supplied with expensive fodder. (Hans-Jurgen Mahnke – The Times, March 4, 1980.)

Farmers now find it more profitable to produce beef for store than for the open market and price guarantees have encouraged many more to enter beef production, creating even greater surpluses. In 1980 over 920 tonnes of beef from Britain and 14,080 from other EEC countries were taken into cold storage each week in an effort to preserve the status quo until 258,113 tonnes were stockpiled at an annual storage cost of £75 million.

Yet at the same time, because of high prices, consumption in Britain is going down and there are fears that the younger generation may already have lost a taste for beef. Some young housewives are apparently so ignorant of the different cuts they are afraid to enter butchers' shops and prefer to buy pre-packed meat from supermarkets or cheaper poultry and processed foods.

Still less are they encouraged to eat beef when they read reports of some of the practices used to promote meat growth. Feed additives, hormone implants, cocktails of vitamins and excessive amounts of anti-biotics can all cause trouble. 'You get strains of disease in animals and then, about three weeks later, you see them in humans,' a biologist told *Now Magazine* in November 1980, cancer, enteritis and septicaemia being among them.

The extreme methods now criticized were begun to fulfil meat demands made by 19th century urban populations. A tempered approach could come in the later stages of the 20th century. 'The writing is on the wall for traditional feeding of concentrates to high yield,' said a farmer in 1979. It is now recognized that Friesian cows may produce more milk than Jersey cows but eat 22% more food, so that in fact the extra milk is produced at an artificial cost. With escalating labour, machinery and bank interest charges, low cost, low return systems are beginning to seem more practical. When in 1980 the first woman championship judge, Mrs Muriel Johnston placed a pure-bred Aberdeen Angus steer called Panda top at the Smithfield Show and a Galloway second, pure beef was put 'back into fashion' and the fact that Panda was bred on an arable farm at Carnoustie in Angus in a 'relatively traditional way' made this seem respectable again. In 1981 a Longhorn, a breed almost unheard of since the 18th century and Robert Bakewell's experiments, won the Burke Trophy, the top beef prize at the Royal Show.

The way forward will perhaps, in part, be the way back. There will be more fields full of richly coloured cattle grown on the land from which they evolved. The cow will revert to its old role of being man's servant but also man's comforter and lead a more natural life. It will cease to be merely a halfway stage between fodder and packet.

The Horse

Although the horse was one of the last animals to be domesticated it probably had a greater effect on the development of the material world than any other. By transporting man and his goods fast and far it enabled him to reach and conquer new lands. In the ancient world horses were scarce and they were reserved for warriors. The drudgery of pulling plough or cart was left to the ox.

Horses evolved during the Pleistocene era; large herds roamed the open plains of Europe, Asia and North America and were hunted by Palaeolithic man for food. By 8000 BC a combination of the spread of forests, after the Ice Age, and men with better hunting skills had succeeded in making the American horse extinct and had forced herds in Europe out of their natural habitat of well-watered grassland and into the arid regions of Central Asia. Fortunately, before the horse was pushed to further environmental limits and almost certain extinction, Neolithic man began husbanding it as a source of food. In about 2000 BC these men, possibly as a result of former successes with cattle, sheep and goats, began herding horses and reintroducing them to their former habitats; chopped horse bones are found among the kitchen remains on many Neolithic sites. Soon after discovering their food value, it is assumed, man discovered the convenience of horses as a means of transport.

Early horses resembled the now extinct Tarpan. They were the size of a large pony, of a dull, smokey-grey colour with a dark eel stripe down their back which ended in a thinly-haired tail. Their stiff, upright manes never grew beyond six to nine inches or fell over down their necks. Their faces were heavy and concave and they had faint zebra markings on their stocky legs. From this horse, it is thought, all domestic ones originated. But because of the huge variation that exists, theories arose, and have still to be

114

thoroughly discredited, that horses evolved from several types. It is most likely, however, that Allen's Law, which laid down the basis of how animals are altered by their environment and climate, affected the horse. In northern climes animals tend to have large, heavy bodies, short legs, small ears and grow long coats in winter, while low lands and hot climates produce animals of finer build with long legs which keep a short, sleek coat throughout the year. Mountain ponies are sure-footed with small, neat hooves; plains ponies have large, wide hooves; and marshland ponies, such as those found in the French Camargue, have wide-splayed hooves. Further adaptations would have been brought about by man selectively breeding for his various requirements.

Quite soon after man took the horse into his control the truly wild horse died out except for those known as Przewalski's horse. These last untamed survivors of the original wild horse of Asia survived under incredible duress on the borders of Mongolia and Russia and only drank at night to preserve themselves from hunters. News of their existence was first brought to Petersburg between 1719 and 1722 by a Scottish official, John Bell of Antermoney. Some other rarely seen white horses existing in a 'cold pole' pocket in the Lena Valley in Siberia may also be wild horses. The reason for thinking they are wild is their single colour. The fact that all other so called wild horses are actually domestic ones who have escaped and gone feral is shown by the mixed colours of their coats. American mustangs and Australian brumbies are of all colours.

Horses, unlike cattle, are not strictly social animals and would have been harder to domesticate. Nevertheless, they are herd animals and operate a highly developed hierarchical system. Stallions occasionally lead the herd, especially during migration, but usually it is the chief mare, and the others follow in order of dominance with the stallion bringing up the rear. Foals follow their mothers in order of birth, the youngest coming first. Young stallions form bachelor groups and live separately from the mares. As they get older, they get restive and make attempts to abduct young mares from other herds. If the younger succeeds in defeating the older he takes his position and his mares, chivvying them into order by nudging and nipping their backsides. He establishes territorial rights over an area, marking the boundaries by urinating and defaecating at selected spots, continually returning to repeat the performance. A horse's feeling for his own range is very strong and is the reason why horses always move faster towards home than away. It is also the reason why all domestic stallions not required for breeding have to be castrated, or stabled on their own, as free-living ones will always try and assert their dominance over other horses.

The catching and taming of horses was probably performed by three separate groups of people (although there could have been a fourth group in Spain) living in the central plains of Asia. Between 3000 and 2500 BC the inhabitants of the Lena Valley adapted to the horse techniques used in harnessing and riding reindeer. For a long time the horse was mainly used as a pack animal or harnessed to light chariots to carry men into war. Horses were unable to draw heavy wagons because of the limitations of early harness. This was borrowed from the ox cart and consisted of a central pole with cross pieces to which a pair of horses were attached by flexible neck collars, which were prevented by a girth round their belly from slipping forwards. When the horse pulled, the whole weight of the vehicle fell on its neck. The collar pressed against the windpipe and jugular vein causing the horse to fling up its head to relieve the pressure and so throw its weight backwards. As a result it was unable to exert its full power and was limited to drawing light vehicles. Not until the rigid collar and traces were invented in about 700 AD could single horses be harnessed to carts. The rigid collar hung too low to interfere with breathing or blood pressure and the horse could throw its full weight on the traces.

By about 2000 BC the people of the Oxus valley were spreading in all directions fighting anyone who crossed their path and using chariots drawn by galloping horses and bronze battle axes. By 500 BC, every community from Persia to India, Asia Minor to Greece and Syria to Egypt was ruled by these aristocrats of war and all those they conquered were frightened of their horses. The vanquished soon learned the art of driving and yoked their own indigenous horses to chariots. The chariot has been described as the 'tank of antiquity' but one with its engine exposed. In the beginning the chariot's speed and the terror it exerted by its strangeness gave its users an advantage but once it was discovered that charioteers could not reach an enemy with weapons, the supremacy was gained by those who fought with swords from the back of horses or with spears on foot. By 1000 BC those great warriors, the Scythians, Bactrians, Huns and Parthians were riding into battle on 'blood sweating, heavenly horses'. The Bactrians had invented polo, the Scythians the stirrup, the Moguls the saddle, and the fame of their beasts reached throughout the civilized world.

Since earliest times certain areas have been famous for their horses. Certain combinations of soil, grass and climate make ideal breeding grounds. Central Arabia with its elevated mountainous limestone hills and flashing streams made wonderful, strong-boned animals and, down the centuries, they and their pedigrees were handed from father to son as an unrivalled heritage. Jealous for their secrets the opposition seized Scythian horses and their riders in battle. Both Greek and Roman bred from Scythian stock. Alexander the Great learned his horsemanship from the Scythians and on his great horse, Bucephalus, conquered two million square miles of the ancient world. Not surprisingly early people worshipped the horse and protected themselves in battle with charms whose relics are seen today in the brasses worn by our remaining cart-horses.

Some of the people displaced by these warring chieftains were a group whose original home was the grasslands surrounding the Caspian Sea. They were known to the ancient world as Keltoi, or Galatoi, and to us as Celts, and they were driven from area after area until their descendants were spread over Europe, Iran and northern India. Their culture was based on the horse and they were perhaps the first people to put the horse to everyday

116

work. They invaded and settle England between 700 and 300 BC but long before this, while Britain was still joined to the Continent, wild horses had wandered into the country. These little ponies, perhaps resembling small Welsh and Scottish ponies were hunted, killed and eaten by Neolithic man and eventually domesticated. The earliest remains unearthed by archaeologists are found in late Neolithic layers at Maiden Castle in Dorset, at Overton Hill, Woodhedge and Thickthorn Down and in Cotswold longbarrows. By the time of the Bronze Age these slender-limbed ponies of about thirteen hands were being harnessed to war chariots and this became their most important function, although they were still eaten and the mares milked.

Because of the mares' short lactation their milk was never as important as that of goats, sheep or cows. The Moguls drank the milk fresh or sour and so did the Tartars. They also made it into an acid-tasting curd called koumiss which was thought to be exceedingly nutritious and an excellent medicine. The Moguls distilled mares' milk into a drink called kurmis and the Tartars into one called rack or racky.

In the late Bronze Age riding was introduced to Britain, and perhaps a more domesticated horse. From the beginning of the Celtic era in 700 BC the horse began to assume something of the major role it was to play in Britain for the next 2000 years. Already white horses were elevated into a cult, indicating their importance as they, in common with other white domestic animals, have always been regarded as special. The white horses on the chalk hills of Wiltshire and Berkshire were originally carved in worship of gods such as Rudiotus, Magog and Epona. To the Celts Epona was the Great Mare or Queen and representatives of her and her foals are found on early sites throughout Europe. Many depict her leading a horse of about 11½ to 12½ hands with fine limbs similar to the Exmoor. This breed is assumed to be directly related to the original wild pony, its ancient lineage betrayed by its broad head, short thick ears and mealy-coloured nose, eyelids and circle round the eye termed a 'toad eye'. These, hardy, stout-hearted little ponies once pulled Celtic war chariots, occasionally carts, but never ploughs. A Celtic ruling laid down that 'Neither horses, mares nor cows may be put to the plough, but oxen in their prime'.

These ponies, harnessed to chariots two and sometimes four abreast, were what confronted the Roman army when it landed on English soil in AD 55. Camden describes in his history *Britannia* how most of the Britons

use chariots in battle. They first scour up and down on every side, throwing their darts; creating disorder among the ranks by the terror of their horses and noise of their chariot wheels. When they have got among the troops of (their enemies') horse, they leap out of the chariots and fight on foot. Meantime the charioteers retire to a little distance from the field, and place themselves in such a manner that if the others be overpowered by the number of the enemy, they may be secure to make good their retreat . . . They frequently retreat on purpose, and after they have drawn men from the main body, leap from their pole, and wage an unequal war on foot.

The British king Cassivellaunus retained 4000 chariots for the purpose of harrying the Romans. Most skilful in the employment of the chariot were the Iceni while Boadicea, with her evil scythed vehicles, mowed down all in her way. However, the chariot had had its day, no matter how skilfully deployed, and the British were doomed to defeat. The onslaughts of the superior Roman cavalry with their superior horses eventually subdued all England. These Roman horses were bred from Scythian stock and were large and fast. Later a heavier, 'bad and ugly' Siwalik German-bred horse was imported along with fresh forces of Roman cavalry. Later both these horses became crossed with British stock. Professor Ewart identified in the remains of a Roman fort at Newstead near Melrose six types of horse. Two similar to Shetlands, one broad-browed and thick-set and another more like a Celtic Exmoor; a larger, coarse-boned Arab type; an even larger Arab 'as fine in head and limbs as modern high-caste Arabs' with skulls like the Jerboa in the British Museum; another resembling the Siwalik horse; and finally a wide variety of cross-breds. To what extent they were used for agriculture or pastoral pursuits is a matter for conjecture. Certainly, horses were now shod for the first time, to protect their hooves against the paved Roman roads. Horse shoes in the modern sense were not invented until the time of Valentinian in 375–92 AD and these early 'hipposandals', invented in Gaul, were usually made of iron or occasionally rope and consisted of flat soles with upturned flanges, and hook and eyelit holes at the back and front for attachment.

When the Romans finally left England they left their excellent horses. Intelligence of the numbers of fast horses in Britain is thought to have reached Saxon raiders. It was not the Saxon habit to fight on horseback but to penetrate deep into the country along rivers and make surprise attacks on foot. Yet once established the Saxons became as concerned with horses and horse-breeding as their predecessors.

The Saxon name for an adult male horse, Steda, is commemorated in names like Stedham, Stead and Steadman. A stod was an Anglo-Saxon mare and Studland in Dorset, Studham and Stoddal in Kent mean pastures where mares grazed. In Norse the word for stallion was Hengest or Hestr resulting in Hestbank in the Lake District, Hestram, an island off Lewis, Hest Fell and Hinksay. Kapall or Capul was the generic name for any type of horse and Capel Crag, Cappelback and Capplerrigg all owe their origin to this.

The Christians abolished the eating of horse flesh as, to them, its consumption was associated with heathen rites and sacrifices. The practice took a long time to die and never did in some areas of the world. The Tartars used to eat horse-flesh raw, tenderizing the muscular parts by cutting them in slices, placing them under their saddles and after 'they have

40. *The traditional plough team: two Shires, a cloth-capped man and a swarm of gulls.*

41. *Horsedrawn sledge in the Lake District, habitually used in hilly roadless areas.*

42. *Welsh ponies 'one of the most beautiful animals' were used for hillfarm work.*

43. Heavy horses taking ammunition to forward guns in the Battle of the Somme, 1916.

44. Dorking Fowl, a breed introduced by the Romans and prized for their fine flesh.

45. The Cockpit, Wm. Hogarth, 1750. Until 1849 cock-fighting was a universal sport.

46. Geese on a Yorkshire hill farm. Today most geese are bred by upland farmers.

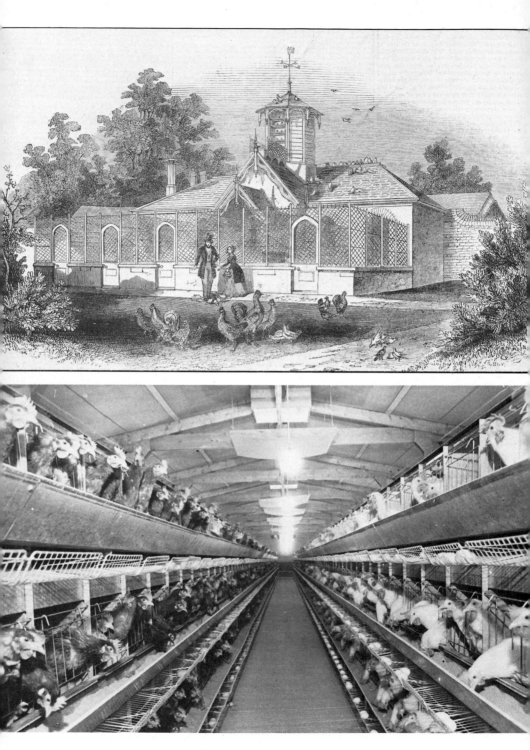

47 & 48. Queen Victoria's Poultryhouse as opposed to a modern intensive unit.

49. *Aylesburys in a Scottish burn; the breed evolved from the Old English White duck.*

50 & 51. The Christmas turkey: Victorian black Norfolk turkeys and modern hybrids.

galloped thirty or forty miles, the meat becomes tender and sodden and fit for the table'. At their feasts the favourite dish was a horse's head but the greatest delicacy was roasted foal. At times of deprivation horse flesh has always been eaten, as it was during the last two world wars. Even today horse flesh is consumed in quantity on the Continent and in Russia.

Since it was no longer eaten in Britain and fighting on foot had become the norm the horse was used in war merely as a pack animal to carry supplies to the front, or as a way of escape in defeat, or a means of transport to the next place of combat. Remnants of the Celtic ponies ridden by warriors in the north can be seen in Shetland and the Western Isles. Here horses were seldom used for ploughing and small ponies were perfectly adequate for the chores of carrying fishing nets and creels from the port and peats from the hill. Each crofter used to keep at least one pony as a mount and for carrying loads. These diminutive black, brown and chestnut animals, often standing no higher than thirty-six inches at the shoulder, possessed immense strength and, trotting with sure feet over the bog and rock of the island, could carry a twelve-stone man or load forty miles in a day. On some of the more isolated islands, such as Eriskay and Barra, unique strains evolved and still survive.

The Norse were expert horsemen. Horses were also valued by the Saxons and their wills are full of bequests of hacks, either broken or feral. In 1015 the aetheling, Aethelstand left a horse Thurbrand had given him and a white one from Leofwine to his father, King Ethelred, a black stallion to the bishop Aelfsige, his horse with harness to his chaplain Aelfwine, a pied stallion to his seneschal Aelfmaer, and a stud at Colungahrycg to his stag huntsman.

Forest studs were common and are frequently mentioned. In 1002 Wulfric gave a monastery at Burton in Staffordshire 100 'wildra horsa' and sixteen 'tame hencgestas' or geldings. The horse was still, like the cow and pig, a semi-wild animal and its breeding was left much to nature. Mares and stallions, when not in use, were turned loose in the forest where they mated and the mare foaled. After foaling she was recaptured but her foal was left to grow up wild and untended until needed for breaking. The custom continued and, to this day, herds of ponies run wild fending for themselves on the hills of Exmoor, Dartmoor and Wales and in the New Forest.

In the 11th century there are hints that horses were occasionally being used for specific tasks on the farm such as harrowing. The first illustration of a horse attached to a harrow is in the Bayeux tapestry.

William the Conqueror's barons and common soldiers came from a country advanced in methods of agriculture and breeding and, added to this, they had the whole of Europe to draw on for their stock. Their cavalry was made up of heavy diluvial horses, some of the many strains of German forest horse, and Spanish horses of 'stately figure and noble action'. Out of William's army of about 12,000 men roughly half were mounted, fighting from the saddle with spear and sword. The Britons still left their horses at the rear and fought on foot armed only with a battle-axe. William's charges won the day, the barons established themselves on the estates they requisitioned, introduced more horses and began to practise the breeding methods they had used at home. Horses began to be graded for use by size. The largest were ridden by knights, slightly smaller ones by ordinary mortals, middle-sized ones usually referred to as stotts or affers operated as pack-horses, small stout animals were used as draught-horses and only the worst were put to work on the land.

The Luttrell Psalter published before 1340 shows these early farm horses to be thick-bodied animals, brown or dappled, shod with pointed shoes and equipped with rope harness. They were kept in stables under the care of a wagoner. He slept beside his charges, washed and combed them and mended their harness. From the early 12th century horses were put to the plough, harnessed in tandem with oxen. By the late 12th century mentions of plough horses increase; at Groton a team is made up of six oxen and two horses and at Somer two plough teams contain the same combination.

From this time on the argument as to whether oxen or horses were the more efficient raged throughout the world until the issue was finally decided by the tractor. Horses might work quicker but they pulled less steadily putting undue strain on primitive plough gear. They were less effective on hard ground, and on the old ill-drained fields where the ox's hoof could spread and give a better hold, but on steep slopes and stoney ground horses were definitely more effective. Horseshoes cost more than ox shoes, and a horse's winter keep was calculated to be four times more than an ox's. They also needed more grooming and were more prone to sickness. Finally, when both were past work, the ox could be fattened on tenpence worth of grass and provide a valuable carcase and hide whereas the horse was virtually useless. As Walter of Henley pungently stated in his *Husbandry* the ox is 'mannes meat when dead, while the horse is carrion'. For a long time, however, the teams remained mixed.

Methods of breeding began to change and become more ordered. The forest clearance that made such a dramatic impact on the lives of swine had an almost equal effect on the horse. In 1086, it is recorded, that there were far fewer 'silvaticae' or wild mares and 'indomitae' or unbroken mares ranging the forests. Furthermore much of what forest remained was enclosed as hunting domains under the Charta Forestae edicts enacted by Henry II. Horses were barred from forest grazing during the winter hunting months and the early summer breeding months and the peasants' rights of 'browse' which enabled them to supplement winter fodder requirements were forfeited. But woods were still the prevailing vegetation, especially in the west and midlands, and here many owners held large herds of feral stock.

Further order was to be seen in the methods of marketing horses. Smithfield is first mentioned in 1171 by William Fitzstephen who notes it as a field for tournaments, races and a horse market and gives a lively description of the scene:

It is a pleasant thing to behold the horses there, all gay and sleek, moving up and down, some on the *amble* and some on the *trot*, which latter pace, although rougher to the rider is better suited to men who bear arms. Here also are colts, yet ignorant of the bridle, which prance and bound and give early signs of spirit and courage. Here also are *managed* or war-horses, of elegant shape, full of fire, and giving every proof of a generous and noble temper. Horses also for the cart, dray, and plough, are to be found here; mares, big with foal, and others with their colts wantonly running by their sides.

Horses for farm work were still considered pretty poor stuff. Sir John Falstolf sent his town horses back to the farm only when they were worn out by work in London. In the accounts of an establishment near Stevenage in Hertfordshire, two horses were kept for carting and eleven for the plough, working in conjunction with five oxen and divided into two teams. In summer cart horses carried oats, hay, wheatsheafs and autumn aftermath from field to barn and in winter firewood, turf, table-wine and food, lead, stone and timber for building and, at all times, water from the river in 'bowges' or leather water-skins. The horses were stalled in winter and kept at grass in summer. Walter of Henley advised housing plough horses for twenty-five weeks from 18 October to 3 May.

Pack horses were mainly used for transporting goods because of the appalling state of the roads. People in carts were frequently thrown out and killed by the violent jolts caused by deep ruts and bumps. The horse you rode was determined by your station in life. Chaucer gives a picture of the range and divisions in the *Canterbury Tales*. Of his 'wel nine and twenty in companye' the Nun's Priest rode a jade, or mare that was past her best; the Monk a palfrey, or comfortable 'thikke', broad-chested riding horse, the Plowman 'rade upon a mare', perhaps a member of his own plough team since mares not stallions were used as they were more docile; the Reeve sat on a 'ful good stot' normally a more expensive horse; the Canon was on a hackney, then a country-bred, inferior, light trotting horse; and the Knight and Squire were mounted on destriers, or great war-horses.

These war-horses had advanced in shape and size along with the design of armour. From 1300 onwards chainmail was gradually displaced by sheets of iron until rider and horse became 'a complete panoply of plates'. These grew heavier and stronger with the increased efficiency of offensive weapons until horses could be called on to bear weights amounting to four hundredweight. This was the motive that Sir Walter Gilbey in *The Shire Horse* says 'impelled our ancestors to develop to the utmost the size and strength of the only breed of horse which could carry a man-at-arms'.

In about 1200, King John whose 'pride' it was to 'render his cavalry . . . as perfect as he could' imported one hundred huge stallions from Flanders, Holland and the Elbe to the 'Lowland and Shire counties'. Many strains of heavy horse owe their origin to the blending of these sires with English mares. Edward I also imported horses to equip his army to fight the Scots.

121

A boost to Scottish stock was given by the capture of these horses after Edward's defeat at Bannockburn in 1314, and from then Edward II's exportation of horses to Scotland was forbidden under heavy penalties. Edward III gave 25,000 florins for fine horses belonging to the Count of Hainault and 1000 marks for Spanish horses to help his army battle against the French. Their power was broken at Poitiers in 1356 and subsequently the export of horses to Europe was prohibited. As a result of these policies English war horses became the superior of any in Europe. This state of affairs was halted by the Wars of the Roses. 'Strong horses' were seized wherever they were found by the contending parties and ordinary men, who saw no future in breeding animals for Yorkist or Lancastrian to purloin, smuggled many of their best animals abroad so, when Henry VII was proclaimed king on Bosworth Field in 1485, the horses he inherited were a fairly poor collection.

Anxious to improve his stock Henry VII passed laws forbidding their export and others prohibiting stallions from being turned out on common pasture. In the 15th century it was the custom, once harvest was in, to turn out all the horses together on common pasture and the consequence was a 'strange admixture' of often inferior animals. As a further control on breeding it became the practice to geld or castrate all indifferent stallions.

Methods of castrating horses have always been fairly barbarous and the ways used today are merely refinements of those employed in the 15th century. They are now, at least, more efficiently applied and the animals are younger (under three months) and their testicles less developed so the process is not as painful. In the 15th century the horse could be anything between five and twelve months and three crude methods were applied. In one the scrotum was opened on each side, the testicles cut off and, to prevent haemorrhaging, the blood-vessels were seared with a hot iron. This has been replaced by the Cutting Method where an incision is made in the scrotal sac and the testes pulled away with the spermatic cord which is not cut as its bleeding can lead to death. A second way was to compress the spermatic cord in a vice between two pieces of wood until either the testicles dropped away or were removed the next day by a gelder who often acted in 'haste, carelessness and brutality'. The so-called Bloodless Castrator method could be said to be its substitute; but rubber rings are now used instead, fitted round the neck of the scrotal sac within the first week of the foal's life and left until they drop off with the testes.

The indignity of the process might be lost on the horse whose intelligence, although rated above a cow's, is below that of a dog or pig, but the pain can be no less real to these extremely sensitive animals. W. Youatt in his book *The Horse* published in 1859 says that the

> development of the brain of the horse should not be lost sight of . . . his perception of music and time are extraordinary, his sense of locality or place is very acute, and the writer never had but one horse that could not find his way back from a new locality better than himself.

In George Orwell's *Animal Farm* 'The two most faithful disciples' to the ethos were the uncomplaining cart-horses, Boxer and Clover. These two

> had great difficulty in thinking anything out for themselves, but having once accepted the pigs as their teachers, they absorbed everything that they were told, and passed it on to the other animals by simple arguments.

The horse's sense of smell and sight are also highly developed; both are products of evolution as they were plain dwellers relying on food growing close to the ground and with no protection against enemies but quickness of apprehension and speed of flight. Thus the horses with the keenest eyesight and smell were the best equipped for sensing danger and so most likely to survive. In the early stages of its development the horse's head tended to lengthen as eyes higher off the ground make better instruments of detection when grazing. The nostrils also lengthened expanding the nasal area which allowed an increase in the thousands of small nerves that make up the sense of smell.

There were still plenty of wild horses in remote forest areas in the 16th century, especially in Scotland. In 1507 the Scottish writer, Boece, mentions that all over the country there were 'gret plente of wild hors'. These ponies gradually became separated by forest clearance and developed into different types all used in the 16th century for everyday work. In Scotland three types of Highland pony evolved; first the small ponies of Shetland, Barra, Rhum and the Outer Isles which were used as pack horses or for drawing transport sledges; second a high-class riding-pony found in the Western Highlands and Isles and now almost extinct; third, the larger Garron, inhabiting Perthshire and the central Highlands, and the strongest of all the pony breeds. Their ancient Celtic origin is shown by their dun colours and a dark eel stripe down the back. In the 19th century they were adapted to carrying sportsmen up-hill and their quarry, which could be a deer weighing eighteen stone, down-hill. The Rev Mr Hall in his *Travels in Scotland* describes their instinctive sure-footedness, saying

> that when these animals come to any boggy piece of ground, they first put their nose to it, and then pat on it in a peculiar way with one of their fore-feet; and from the sound and feel of the ground they know whether it will bear them.

Further south in Galloway, Gervase Markham says in 1660, there were a 'race of small nagges which they call Galloways or galloway nagges which for fine shape, easie pace, pure metall and infinit toughness are not short of the best nagges that are bred in any country whatsoever'. The breed began to die when a law was passed by James I in 1605 depriving the 'moss-troopers and other predatory border men of a method of livelihood which involved the use of hardy and enduring horses'. True Galloways now no longer exist but before 1800 Gilbey says in his *Thoroughbred and Other Ponies* (1903) these dark-coloured animals of about 14 hands, played an active part in agriculture work in the lowlands of Scotland. In localities where no roads existed, and wheeled traffic was impossible, galloways were used not only for riding but for the transport of agricultural produce.

Across the border in Cumbria similar ponies, probably relics of the horses used by the Romans stationed in that area, were also used for general-purpose work. These are now classed as Dale and Fell. Fell ponies were formerly called Brough Hill ponies after an annual fair held at Brough Hill in Westmorland. They were much in demand by packmen and the Fell-side farmers became adept at breeding fast-trotting small cobs.

In the mountains of Wales there were more small, hardy Celtic ponies and in the lower valleys a larger Cob type. A writer to a livestock journal describes them as 'very compact, and with remarkable pluck, being able to move heavy weights in the hilly country of Wales which would have been thought beyond their power'. Pony-hunting with lasoos used to be a favourite sport of the Welsh farmer and peasant. The captured beasts were sold at Bala Fair to supplement their meagre incomes. The ponies also pulled harrows which were attached to their tails.

As the forests that once stretched in a mass across the whole of southern England became concentrated into one area, the New Forest, another breed evolved. New Forest ponies are described by Youatt as being 'generally ill-made, large-headed, short-necked, and ragged-hipped; but hardy, safe and useful, with much of their ancient spirit and speed'. The introduction of Arab stallions to the forest in the 19th century did much to improve their looks but not their hardiness.

On Exmoor there were ponies which 'although generally ugly enough, are hardy and useful'. And on Dartmoor there was another 'race of ponies much in request in that vicinity, being sure-footed and hardy, and admirably calculated to scramble over the rough roads and dreary wilds of that mountainous district. The Dartmoor pony is larger than the Exmoor, and if possible uglier'. Both were used for everyday work on the farm, pulling sledges and acting as pack horses. Further west on the Goonhilly Downs in Cornwall there was a tiny pony that gave the generic name of Goonhillies to all West Country pack and riding-horses.

Goonhillies and many another small pony disappeared when Henry VIII decided to continue the work of his father and enacted legislation designed to improve further the general strength and stature of horses. In 1535 worried that

in many and most places of this Realm, commonly little Horses and Nags of small stature and value be suffered to depasture, and also to cover Mares and Felys of very small stature, by reason whereof the Breed of good strong Horses of this Realm is now lately diminished, altered, and decayed.

he laid down that all owners of enclosed grounds a mile in compass should keep two mares able to bear foals of thirteen hands and they were to be covered by stallions of at least fourteen hands. In 1541 he introduced more stringent rules for the Shires dictating that where mares and fillies were kept no 'Stoned horse above the age of two years, not being fifteen hands' should be 'put in any forest, chase, moor, heath, common, or waste.' Every archbishop and duke was summarily ordered to keep seven trotting stallions of fourteen hands for the saddle, and 'every layman', whose wife wore French hoods or velvet bonnets, one such stallion. Henry also strengthened the law prohibiting the export of stallions and mares worth more than six and eightpence, imposing a lashing fine of forty pounds, and procured for breeding the finest animals that 'Turkey, Naples, Spain and Flanders could produce'.

The husbandry of horses was also improved. Those destined for war or transport were pampered compared to other farm stock. They were fed throughout the year, except during August and September, on hay, beans, malt, oats, wheat and barley-chaff, peas and vetches. Lean horses were fattened on wheat and barley and given a mash of ale and wine. Cart-horses, Gervase Markham advised, should be fed sweet hay, chaff, peas and oat-hulls and the Carter or horse-keeper should be patient and use 'fierce words' more than 'stripes'. He was to rub and comb his charges every morning, wash their feet after work with water, wine or ale and butter and to rub their legs with butter and oil to strengthen and supple the sinews. Horses were to be placed in the stable according to their liking for each other and the stalls must be cleaned every morning. Carters had to sleep with their horses in case they broke loose and fought and each must 'look advisedly and warily to his candle' to prevent fire. To promote good health in the horses many suggested bleeding them on St Stephen's Day, 26th December, and feeding them well afterwards.

St Stephen was the patron saint of horses and the bleeding of animals might well have been connected with an ancient idea that it symbolized his martyrdom.

By the late 16th century the implementation of all the rules of breeding and husbandry began to tell, for Harrison states that pack horses could carry four hundredweight 'without any hurt or hindrance'. Thomas Blunde-ville in 1565–6 admiringly described the Great Horse as

> not finelie yet stronglie made he is of great stature. The mares also be of a great stature; strong, long, large, fayre and fruitful; and besides that, will endure great labour in their wagons, in which I have seene two or three mares go lightly away with such a burthen as is almost uncredible.

Gervase Markham said that horses for carts and ploughs should be of average height, in order that they could be easily matched, 'of good strong proportion, bigge brested, large bodied, and strong limbed, by nature rather inclined to crave the whipe than to draw more than is needful'.

But the end for which all these huge animals were bred was disappearing.

The age of gun-powder and firearms was arriving and no amount of muscle or armour could withstand their onslaught. Lighter, swifter horses to carry cavalry were demanded and 'running horses' for the new sport of racing. However, a new role was immediately found for any heavy-horses spared by the army. Carriages were introduced in 1580, supposedly by the Earl of Arundel, and were quickly adopted by Elizabeth I. The fashion spread so fast that every heavy-horse was 'rapidly brought up for the purpose, and became so exhorbitantly dear, that it was agitated in Parliament whether the use of carriages should not be confined to the higher classes'. When Queen Elizabeth moved from one place to another the historian Ralph Holinshed commented that 'there are vsuallie 400 carewares, which amount to the summe of 2,400 horses appointed out of the countries adioining, whereby her carriage is conveied vnto the appointed place'.

In order to pull the carriages through the deeply-rutted roads and mud of the late 16th century horses of supreme strength were required. The roads continued to be bad for the next two centuries. In 1770 Arthur Young measured ruts 'four feet deep'. On the turnpike from Newstale he was 'obliged to hire two men at one place to support my chaise from overturning' and found the road from Chepstow to Newport 'full of hugeous stones, as big as one's horse, and abominable holes'. The principal road from Tamworth to Ashby was 'in a state almost impassable several months of the year' and waggons had to be taken off their wheels and dragged on their bellies.

Because of the requirements of carriages at first only 'misfits' from the army found their way into the farmyard. But when the Civil War ended in 1647 armour became obsolete and the Great Horse, no longer needed for service, flooded into civil life and from there into agriculture. Cromwell's victory over the Royalists was partly due to the fact that he drew his forces from the midlands and East Anglia, areas which traditionally bred the finest heavy-horses and, moreover, he had controlled the east coast ports through which the excellent black Friesian horses were shipped from the Low Countries. It was in these areas that the 'proud but of necessity somewhat lumbering war-horse set to draw coaches through the mire of execrable roads' first descended to cart and plough and it was from here that the main breeds of heavy-horse emerged.

Throughout the 17th century breeders turned their attention to producing better horses. Once again East Anglia, and particularly Norfolk, had a head-start over other regions since it was the home of the great agricultural improver, Thomas Coke. He decided to set an example by showing how much more efficiently and quickly two horses and a single man could plough than 'six oxen attended by a man and a boy'. Coke and other farmers were further aided by the drainage of the Fens in the late 17th century and the rich agricultural land it exposed. And when this was enclosed, instead of the fields being laid out to suit oxen, they were blocked out in areas fitted to working horse teams. Furthermore there was already a basic horse suitable for pulling ploughs across this sodden land. These Suffolk horses valued for 'burden or draught' were, according to Arthur Young generally about fifteen hands high, of a remarkably short and compact make; their legs bony and their shoulders loaded with flesh. Their colour is often of a light sorrel . . . for draught they are perhaps unrivalled . . . An acre of our strong wheat land ploughed by a pair of them in one day, and that not an unusual task, is an achievement that bespeaks their worth, and which is scarcely credited in many other countries.

Young was not all praise for Suffolks, accusing them of being very plain-looking and of being able only 'to walk and draw; they could trot no better than a cow. But their drawing power was very considerable.' This is attributed to the low position of the Suffolk's shoulders which allows the horse to throw all its weight into the collar. Draught power is closely allied to weight and is aided by proper harness. The collar should lie evenly and flatly upon the collar-bone and shoulder muscles, to allow them freedom of action, and all the traces should be as near horizontal as possible to the carriage for ease of pull. A Suffolk stallion can weigh as much as twenty-two hundredweight and round Woodbridge, Debenham, Eye, Lowestoft and Orford, where the best horses were found, they were 'all taught with very great care to draw in concert'. Keith Chivers in his book *The Shire Horse* (1976) includes a quote from an early writer who said:

many farmers are so attentive to this point, that they have teams, every horse of which will fall on his knees at the word of command twenty times running, in the full drawing attitude, and all at the same moment, but without exerting any strength till a variation in the word orders them to give all their strength – then they will carry out amazing weights . . . I was assured by many people here, that four good horses in a narrow-wheeled waggon would, without any hurt or mischief from over working, carry 30 sacks of wheat, each of 4 bushels 30 miles, if proper fair time was given them.

Drawing matches to test the best became famous. Such is the heart and willingness of the Suffolk that quite a few ruined themselves by their exertions.

Some of the Suffolk's power came from its capacious belly – the barrel that gave the 'Punch' to the Suffolk's name. This enabled it to carry its food longer than other horses and stand a ploughing day that began at 6.30 in the morning and ended without a break at 2.30 in the afternoon, so giving Norfolk and Suffolk farmers a reputation for ploughing more land in a day than any others in the island. Much of the Suffolk's 'Punch' was bred out of it in the 19th century by breeders trying out new theories and also attempting to adapt the horse to the carriage. Today the Suffolk is a lighter, thinner breed standing about 16 hands. However, its colour is still distinguished by the old name of 'sorrel' and perpetuated by many a Suffolk pub sign. The colour is really chestnut and sometimes the face is decorated by a white star or blaze.

127

Across the waters of the Wash, in Lincolnshire, another heavy horse was bred from the old war-horse. This animal was generally black and its speed, no longer needed to charge into battle, had slowed down to such an extent that Marshall referred to him as 'the black snail breed'. Another similar, but more mettlesome, black horse was bred in the midlands and, in the 18th century, both were crossed and mixed with imported Friesland and Zeeland blood with the aim of making an animal suitable for work in the newly-emerging town docks and industries. Heavy-horses of the 'highest perfection' now began to take shape.

Several people had a hand in their production. Lord Chesterfield, during his period at the embassy in the Hague which began in 1728, sent six Zeeland mares back to his estate at Bretby in Derbyshire for breeding. The Earl of Huntingdon also 'brought home with him a set of coach-horses of the black breed from the Continent', all stallions, and used them to mate with local mares. Finally, Robert Bakewell, anxious to try his breeding methods on horses as well as sheep and cattle 'made a journey through Holland and part of Flanders, and there purchased some West Friesland mares, which excelled in those parts wherein he thought his own horses defective'. These sturdy, jet-black Friesland horses, anciently used for riding, war and agriculture, with their high-arched necks and high-stepping gaits slightly resemble Fell ponies in looks but are much larger.

Bakewell put the Friesland mares he brought to his own Leicestershire Black Stallion, breeding in-and-in to fix his desired type. The horse that finally resulted from all these experiments became known as the Shire horse, named after the Shire counties of the midlands. It was, and is, the largest draught-horse, standing between 16 and 18 hands. It has a coarse, rather heavy head, a short body of immense strength, can be black, grey or chestnut in colour with white socks, spectacularly feathered with soft hair, and was known as a steady worker that did not shirk its duties.

While the Suffolk and Shire were being evolved in England, the Clydesdale was based in Scotland on the foundation of the Old Scottish War Horse. Some of the Clydesdale's blood came from war-horses left floundering in the mud at Bannockburn by the defeated English but more came from successive imports, over the centuries, of breeding horses from Hungary, Poland and England. By the 18th century these had produced in Upper Clydesdale a heavy horse with a broad head, a large ear, a lively eye and action and a hardy constitution. This horse was further improved by the importation of a 'black Flemish stallion' from England by John Lochlyoch, 'four well-sized English mares with foals at foot, one black stallion as well as other colts and fillies' by Johnstone, Laird of Alva and 'a remarkable strong black horse' acquired by the Duke of Hamilton from Robert Bakewell. Some Cleveland Bay blood from Yorkshire was also probably used and the result, known as the Lanarkshire breed, became the common horse of the Scottish Lowlands for, according to Ure in his report on Roxburghshire, 'They draw more surely, and are better for heavy work in the field, than any other.' All over the counties north of the border, these horses were bred, and then sold to farmers in Renfrew and Ayr for breaking and training. At five years of age they were sold again at fairs at Rutherglen and Glasgow for work in eastern Scotland and the north of England. Clydesdale horses, as they came to be called, stand between 16 and 17 hands and are bay, brown, black or roan in colour. They have white socks on long legs and a narrower body and more nervous temperament than other heavy horses.

The blood of these three great cart horse breeds, the Suffolk, Shire and Clydesdale, was disseminated throughout the country by several means. Robert Bakewell pioneered the practice of letting stallions for a season to a farmer or group of farmers. Thus, Marshall eulogized,

> even one superior male may change considerably the breed of a country. But, in a year or two, his offspring are employed in forwarding the improvement. Such of his sons as prove of superior quality are let out in a similar way; consequently the blood, in a short time, circulates through every part, and every man partakes of the advantage.

Others kept their most valuable stallions at home and mares were sent in for mating. The ordinary farmer could not afford the time or money to send his mare away and took advantage of 'travelling stallions'. These were led round a district stopping at points, which were advertised, to receive the mares of the neighbourhood. A busy stallion could have the following schedule;

> To Cover This Season
> At William Spinke's, Eyke
> A Bright Chestnut CART HORSE at 15s. 6d. each Mare.To pay the Man at the time of Covering. He is 16 hands 1 inch high, short leg'd and large boned. He will be at Alderton Swan on Tuesday, at Bealings Admiral's Head on Wednesday mornings, from 9 to 11; at Woodbridge White Horse same day; and at home that night; at Saxmundham Bell on Thursdays; and Bramfield Queen's Head on Fridays; at Framlingham Crown on Saturdays; at Parham-Hacheston Queen's Head that night; and at home on Mondays. Such mares as were not stinted last season may be covered at 8s. 6d. each.'

The disadvantage of this system was that you could pick a dud stallion. Another difficulty with travelling stallions was that their time in your area might not coincide with your mare's heat period. Mares come into a weak 'foal heat' seven to nine days after giving birth and then again three weeks later. A horse's desire depends on daylight. Their breeding period naturally begins in spring and their sexuality intensifies with the intensity and length of daylight hours. Gervase Markham in his *The Perfect Horseman* written in the 16th century gives a graphic description of how

> to know when your Mares are ready (if it be in a wild State) observe their chasing and galloping up and down morning and evening, and

their inconstancy of abiding in any one place, especially throwing their noses to the North and South, the lifting up of their tails, riding one another's backs, wooding one another, oft pissing, and opening of their shares and closing them again, are all signs of Lust.

Breeders often use a 'teaser' to test and raise a mare's heat in order to shorten the time a valuable stallion has to spend with her. A more romantic approach is advised by Markham who suggests putting

them together into some close-walled Paddock, where there is store of sweet grass and sweet water, just upon the going down of the sun, as near as you can observe, either three days after the change, or three days before the full of the Moon, and let them remain close together two whole nights and one day, and take the Horse from her at sunrise.

In their wise old way these writers knew more from commonsense and observation about what was right. Circling the mare with his head held low, nipping her hindquarters and false mountings are all the stallion's ways of stimulating the rise of his penis, which is slow because of its peculiar type of muscular tissue. Once the penis is erect the stallion immediately 'leaps the mare', and thrusts like a human for several minutes until the emission of his sperm. Then he rests on her back for a while before withdrawing and dismounting and, if left with the mare, will re-serve her five to ten times during her heat period. Conception complete the mare gives birth eleven months later.

Mares were worked almost up to foaling, and only given a month off work afterwards. Foals were shut up during working hours and those born in March and April were weaned in October or November. Some were sold with their dams at the autumn fairs of Ashby, Market Harborough and Loughborough, but most were kept until they were eighteen months old and dispersed at Burton-on-Trent, Rugby and Ashbourne. These fairs were great gathering places where the pleasure often equalled the business transacted. Absurd competitions such as 'sneering through a Horse-collar' and 'drinking the Hottest Cup of Tea' took place while horse races and dances created other diversions.

The foals were bought by midland graziers who ran them for a year with their other stock before selling them to arable farmers at Stafford and Rugby who broke them to harness. First the colts were haltered, and then 'gentled to lead', walk and turn about. At two or three years old they were put in full harness, stood in a stall and walked about the yard to get accustomed to its feel. Then the colt was placed in a plough beside an experienced horse and worked for half-days only. If a horse was 'well-made' by sympathetic hands it could be controlled by 'a thread' and guided without effort. They were then worked in teams until five or six years old. These teams were often huge. Youatt explains the reason, saying,

The farmer is training them for their future destiny. and he does right in not requiring the exertion of all their strength, for their bones are not yet perfectly formed nor their joints knit; and were he to urge them too severely, he would probably injure and deform them.

A town was the next stage, and most horses went to London: bought by dealers for use in 'drays, carts, wagons, coaches, the Army, or any other purpose they turn out to be fit for'. When worn out by the streets, these horses returned to the land, frequently being bought by small farmers changing over from ox to horse. A cheap way of doing this was to buy cast-offs from town.

In the 17th century 'huge machines looking at a distance more like a cart than a plough with a beam the size of a gate post' were the norm. Early wagons weighed over a ton, had wheels six inches broad, and measured six feet wide and fourteen to fifteen feet long. Improvements were made and by the end of the 18th century lighter, better-designed, two-wheeled wagons made their appearance and a swing-plough able to be manipulated by two horses of the new breeds. The ox team retreated in the face of this progress but slowly. In Kent 'four, six, yea twelve horses and oxen' moved cumbersome ploughs while in progressive Norfolk two horses pulled an innovatory double-furrow plough. However, gradually everyone saw that 'a pair of mares and one man will do as much work as four oxen and two men'. But in the West Country, outlying parts of the midlands and southern counties the 'groaning ox' was doomed to labour on. The backwardness of many communities was partly due to difficulties of communication. In many regions there were still no roads and sledges continued to be used for moving goods. The roads that did exist remained appalling and could only be traversed in places by ponies. They were used not only for riding and on the farm but also by packmen transporting goods from town to town. Sometimes the packmen led a single horse and sometimes a string, loaded down with goods slung over their back in panniers. In the north-west they also carried salt from seaside salt pans or the inland boiling centres in Worcestershire and Cheshire. Vast quantities of salt were required for preserving meat especially for victualling the army and navy.

The people of the North Riding of Yorkshire divided their horses into two types; a galloway 'nag', or saddle-horse that would also go in harness, and the 'chapman' pack-horse, a short-legged animal. This horse developed, along with the advent of tarmacadam and the need for lighter, faster horses to pull post-chaises and new, slimmed-down coaches, into the Cleveland Bay. Breeders added Arab blood to 'combine action with strength' so cleverly that 'Cleveland and the Vale of Pickering in the East Riding of Yorkshire, may be considered as the most decided breeding counties in England for coach horses, hunters and hackneys'. This bay or brown horse standing 16 hands with a proud air has a large head and a strong, lengthy body carried on short, clean legs. It now pulls Royal coaches, is used in driving competitions and has won Olympic medals for show-jumping.

As tarmacadam improved the highways and travellers could journey more smoothly, the ambling roadster, with its easy pace, was supplanted by the hack, and the cumbersome great horse and pack pony gave way to the coach and wagon-horse. These in turn were ousted by the railway. Dray-horses were still required in towns but apart from this, increasingly, the only place for the heavy horse was the farm.

For the next hundred years teams of heavy horses gallantly pulled plough or harrow and prepared the rich acres of Britain for seed to feed the people of the boom years of the Industrial Revolution. The common picture most people have of fields being ploughed by teams of gigantic horses straining at their collars and directed by a cloth-capped man surrounded by a swarm of wheeling gulls seeking worms in the newly-turned tilth is, in fact, of reasonably recent origin and lasted for a relatively short time. But during this period the breeding of the Clydesdale in lowland Scotland and the north of England, the Shire in the midlands and the Suffolk Punch in East Anglia reached heights of excellence and uniformity of type. Their improvement was helped by the establishment of local clubs and societies who controlled the hiring of stallions and later by regular classes for horses at agricultural shows. George Borrow says that at Tombland Fair, near Norwich the

> goodliest sight of all – certain enormous quadrupeds only seen to perfection in our native isle; led about by dapper grooms; their manes ribbanded and their tails curiously clubbed and balled. Ha! ha! How distinctly do they say, Ha! ha!

At these shows sires could be exhibited and judged in comparison with others. The combination of their best points was fixed in writing by the Breed Societies that all came into being during the 19th century.

Word of the excellence of British horses spread. Once Europe had recovered from the disruptions caused by the Napoleonic Wars vast numbers of horses began to go abroad. It took 750 years of importations from Europe to make the English heavy-horse but now the shoe was on the other foot and the world wanted our breeding stock. Huge prices were offered and gratefully accepted because of a depression in farming at home. Good horses became a rare commodity in Britain and practically unsaleable here. And when farming conditions improved there were no suitable horses to work the land. Yet the horse had become a farming necessity. Mowers, reapers and other farm machinery were now all mechanical and designed to be drawn by horses. In an effort to keep pace, blood was brought in from the continent and horses mated indiscriminately. The feeble mongrels that resulted did no one any good. Enlightened landlords began to take a hand, breeding fine stallions and lending out their services to tenants. Their policies had an effect and once again superlative horses were produced.

The formation of the English Cart Horse Society in 1878 opened a new era in horse history. 'Real worth in horseflesh', said its chairman Sir Walter Gilbey, 'is never put out of service'. Breeders from Royalty down to tenant-farmers concentrated their energies on producing fine examples of these 'Great gallant lion-hearted workers . . . friendly as kittens'. Shires of repute came from studs owned by Edward VII, the Dukes of Devonshire, Westminster, Bedford and Sutherland, the Lords Rothschild, Bentinck and Wantage, Sir Philip Muntz and Sir Walter Gilbey. Many of the greatest were bred in the Ashbourne district of Derbyshire where the limestone land helped imbue the Shire with all the qualities that make a great horse.

After the formation of the Cart Horse Society it became the custom to hold your own horse sales. In February, 1885 over 2000 people travelled to Sir Walter Gilbey's sale at Elsenham. The horses fetched record prices for the time. An average of £172 was paid for each and the total sale realized 6,561 guineas. The record price ever was reached in 1913 when 4,100 guineas was paid for a dark brown two year old, Champion's Goalkeeper at the dispersal sale of Lord Rothschild's Tring Stud.

The Clydesdale also had its 19th century champions. Its background is thick with the names of the landed gentry of lowland and central Scotland; Stirling of Keirs, Somervilles of Lampit, Muirs of Bowhouse and Weirs of Sandilands all laid secure foundations for the breed. The Duke of Hamilton made a lasting impact when he commissioned Lawrence Drew to buy horses at Battersea International Show in 1862. Drew chose two, a stallion called Sir Walter Scott and a mare named Maggie, and with these established a stud which eventually bred the Prince of Wales. This famous sire left his mark on the whole Clydesdale breed. Lawrence Drew's breeding methods became discredited when, in an effort to put weight on the Clydesdale, he bred in too much Shire blood and, because of the influence of Drew's stud, this blood affected the whole breed and for a while its reputation fell. Farmers bred back to type and the Clydesdale's fortunes recovered.

Five main 'tribes' influenced the Suffolk breed; the Blake tribe which emanated from Farmer, a stallion owned by Andrew Blake; the Allenborough tribe from Farmer's Glory bred by John Wright; Barber's Proctor was the basis of the Shadingfield tribe; Samson 324 was the father of the Samson tribe; and last, and most important, The Horse of Ufford produced by Thomas Crisp's tribe. From the seed of this bright chestnut, foaled in 1768 and advertised five years later as 'able to get good stock for coach or road', almost every Suffolk in existence can claim direct descent.

'These three heavy-horses' Robert Trow-Smith claims in his *British Livestock Husbandry* 'offered the British farmer of the draught-horse era all the choice he needed: indeed any one of them could have met the whole national need'. A distinct way of life grew up round the plough horse; George Ewart Evans gives a complete picture of the one that evolved round the Suffolk Punch in *The Horse and the Furrow* (1960) and this is summarized below as the customs that existed there could, with few exceptions, be translated to the whole country.

The ploughman's day began at 4.00 a.m. when he took a bite of bread and cheese and hurried to feed the horses for between the time the horses had their first *bait* or meal and their turning out to plough at 6.30 a.m. two hours must elapse. This was an unalterable rule in Suffolk; to give the horses less time was to treat them unfairly since they had nothing more to eat until returning to the stable at 2.30 in the afternoon. It was essential, therefore, that the horse should have a good morning meal and plenty of time for digestion. These hours were kept throughout the spring and summer but from October to April they turned out half an hour later, at 7 a.m. On the Battisford estate there were twenty-four horses – eight plough teams – with two horsemen and two *mates*, or under-horsemen, to look after them. The order of precedence of the horse-men was fixed with almost military precision; and at no time was this more jealously guarded than when the teams turned out in the morning and returned to stable after work. The horse-men and their mates went first with their teams and if, on a particular morning, more teams were needed they went in the charge of *day-men*, or ordinary farm-workers and followed on behind.

In the middle of the day, at eleven o'clock, the teams stopped working. The horse-man threw a couple of sacks over the backs of his horses and sat under a hedge to eat his *elevenses*. The break lasted twenty minutes; but it was not a complete break as the horseman was still in charge of the horses. Occasionally these would become restless (especially in summer time when flies were bad) and could get a foot over the trace. After the break work resumed and continued until 2.30 when it was the end of the day as far as the horses were concerned.

The same order of precedence was observed at the end of the day and, if a second horseman happened to finish ploughing somewhat near the gate, he had to 'hold to one side' and draw his horses away until the first horseman and his mate had passed through. The horses walked back to the stable on the right-hand side of the road and, if a horseman rode, he sat sideways on the *land* horse, the one on the outside nearest to the oncoming traffic. As soon as the teams had returned to the stable the horses were uncollared and given a handful or two of stover. The baiters then returned home for their dinner. At 3.30 they came back to the stable and for two hours fed the horses and removed the mud and dust of the day. The horses were never allowed to stay in the stable at night but, at about 8 o'clock, were turned out into a well-littered straw yard with plenty of good sweet oat or barley straw to eat. With this treatment the horse's legs never swelled and he never 'had any ailment about him'.

It was a matter of pride for the baiters to turn out their horses looking as fine as possible since they did not like the horses on neighbouring farms to have more *shine* or *bloom* on their coats. Tansy leaves, bryony root, black antimony and chaff wetted with urine were all sprinkled in their feed and the horses' coats were lightly rubbed down with rags dipped in paraffin. Harness and brasses were brilliantly polished and, to add colour to the scene and ward off flies, straw hats were worn by the horses at harvest time.

Fields were ploughed in stretches of varying widths, each bordered by 'water-cuts' or deep furrows for drainage. The lighter and better-drained the soil the fewer water-furrows were needed and the wider were the ploughed stretches. All cultivation was done along the stretch itself, which meant that every implement whether drill, hoe or harrow, was adapted to fit the width of stretch used in that locality. To begin, the ploughman first made an immaculately straight furrow, guided by a 'stetch-pole' or peeled hazel stick placed at one end. After the first furrow had been drawn, a second was laid alongside to complete the 'laying of the top', rather like the ridge of a roof. A level top, apart from looking good, enabled the drill coulters to bite to a uniform depth and sow the seed evenly. The ears of corn would then mature at about the same time.

Land was ploughed not once but several times throughout the winter. Directly after harvest the earth was turned for the first time as William Cobbold explains;

The *first earth* was the first ploughing. For the *second earth* you turned back the furrows, ploughed them back until you were as you were before. For the *third earth* you crossed it: that means you ploughed it across or *overwart* as we called it: you ploughed at right angles to the first furrows made. After the *third earth* you let the land lie for a while. It would be very rough, great old clods of soil and stuff but the elements would get to work on it and help pull it down. You then harrowed and rolled it and kept it moving. Most of the old spear-grass would come to the surface after the second ploughing: the harrowings would bring it out for you to burn. At the *four earth* you ploughed it back – turned it so you were *as you were* once again. Then you came to the *five earth* – the really skilled part of the whole job – when you stretched up the land for your crop of corn.

Next the land was mucked from a tumbril, this tipped and the manure could be raked straight from the back and onto the land. The tumbril's wheels were set exactly 62 inches apart so that they travelled in the outside group of a set of three furrows measuring 31 inches. The horse moved down the centre furrow so that neither hoof nor wheel upset the work. The ridges were then split and the soil thrown back to cover the muck. After this they were rolled, then drilled and seeded. When the corn was two inches high it was hoed and finally, when ripe, reaped.

The boom in 19th century agriculture ended in 1874 and, in the subsequent depression, the export of horses was frequently the farmer's only source of revenue. America was prepared to take every English-bred horse available to pull their ploughs across the grain fields of the middle west. The demand became so great that the unscrupulous exploited the market. The Americans' own depression in 1890 put a stop to the flow and, when it resumed, ten years later they had learned from the experience and went for quality not quantity. Exports began again and rose from 159 in 1900 to 504 in 1910 but then dropped sharply to 185 in 1913.

There were two reasons for the decline; first was the British fashion for excessive feathering on the Shire's legs which might look good on a display horse but was inconvenient on a working horse, acting as a collecting centre for dirt and mud; and second was the foreign competition as the Americans found they could buy French and Belgian horses for less than English. In spite of this, top horses continued to go to the States and to the developing countries of Canada, Australia and New Zealand. Shire, Clydesdale and Suffolk all moved the virgin tilth of these countries.

Then came the war. The Great Horses that were once required to carry the weight of knight and armour, were now needed to draw massive guns. In order to fight the Germans we needed every heavy-horse we owned. 140,000 were compulsorily purchased by the army in 12 days in 1914, the horses added to their existing force of 25,000. By the end of the war more than a third of all the horses in the country had left for service and by the time the fighting was finished 500,000 had died, along with 1,000,000 men, in the battles of the Marne, Mons, Ypres, Vimy Ridge and Passchendaele. At home, the land was denuded not only of horses but of essential craftsmen like blacksmiths, wheelwrights, harness and implement-makers.

The farm-horse world was shattered by these blows. To help reorganize the industry a Food Production Department was established in 1917, to act as a clearing-house for the needs of individual farmers, and a Cultivation Branch organized the manufacture or importation of ploughs, reapers, binders, harrows and other implements. It also re-trained ploughmen and operated 10,000 horses but, to the ire of the breeders of the Shire, Clydesdale and Suffolk, they were foreign horses. During the war British Remount Purchasing Officers, desperate for supplies, brought Percherons from America. They were very impressed by the horses' activity, hardiness, good temper and fortitude during long journeys by sea and rail and under the fire of battle. Percheron stallions were also able to beget uniform progeny from indifferent mares and in 1916 Lord Lonsdale was sent to France to buy six mares and a stallion of this originally French breed.

The Percheron is more closely related to the mediaeval war-horse than any other. Unlike English horses this stocky, grey-dappled animal never had its charging speed bred out of it; it remained a faster horse, its short 16 hands body packed with vigour and strong bone with sound hard, blue hooves developed by the rough, stone roads of France. The Percheron was for a long time the working horse of northern France and famous for its speed, healthiness and even temperament. Arab blood, bred into it at one point, gave it a dished face, refined the animal and caused variations in the breed which still exist.

A Percheron society was formed in Britain in 1918 much to the chagrin of other heavy-horse breeders. It was partly their own fault for breeding British heavy-horses too much for the show-ring. 'The proof of a good horse is what he will do when he gets his collar on not how he stands in the show-ring,' said a breeder and the government, in agreement, instigated a Heavy Horse Breeding Scheme to try and rectify the situation. This act controlled the type of stallion 'Travelling for hire or exhibited in public places or markets or shows' and was aimed not only at improving the general standard but also at increasing the number of horses since the Government expected that, after the war, many would be needed both by the army and civilians.

The government's anticipations were wrong. When in 1921 the foals bred under their impetus reached maturity there was no need for them in civilian life as the 'internal combustion engine' had arrived. Neither was there any need for army horses as the services became motorized (and not only stopped buying new animals but sold those they already had). The government abandoned their Heavy-Horse Scheme and prices crashed. The warning rumblings had been there for some time. In the 19th century steam engines, although too unwieldy to succeed on farms, had nevertheless been tried, making people aware of their possibilities. Before the First World War the car had already challenged the light-horse's position on the road and from there it was only a short step for the tractor to supplant the heavy-horse on the farm. Even so it took another twenty-five to thirty years generally for the average farmer to abandon horses.

During the 1920s the practical shortcomings of the tractor restricted its use. However, tractors improved, became cheaper and the extra work they could perform at busy periods, coupled with the rising cost of labour, persuaded even the most die-hard of their advantages. Yet there were arguments against them. A tractor might initially cost the same as two horses and their gear but its working life was only four or five years whereas a horse's was ten and it could then be sold for slaughter. Horses might have to be fed at idle times of year but then tractors depreciated and rusted away. Horses provided employment which, during the depression of the 1930s was a valid point, and their appetites provided a market for produce since it required three acres to feed a horse and, on a mixed farm, feeding was cheap whereas petrol or diesel oil were expensive. On a small farm a horse was more versatile and it was on these farms that the horse lingered on longest.

Laurie Lee in his idyll of Cotswold life, *Cider with Rosie*, mourns the passing of the way of life that went with these horses, saying:

Myself, my family, my generation, were born in a world of silence . . . of white narrow roads, rutted by hooves and cartwheels, innocent of oil or petrol, down which people passed rarely, and almost never for pleasure, and the horse was the fastest thing moving. Man and horse were all the power we had – abetted by levers and pulleys. But the horse was king, and almost everything grew round him: fodder, smithies, stables, paddocks, distances, and the rhythm of our days.

The last census of horses in Britain was taken in May 1934 revealing that in ten years total numbers had decreased from 1,892,205 to 1,278,341. Of this number agricultural horses had fallen the least, from 754,000 to 653,000. Then in 1936 the government began to plan the logistics of feeding the population should there be a war. Mechanical power, it became evident, would have to be employed to plough and cultivate the two million extra acres it was thought would be needed. In April 1939 the Minister of Agriculture, Sir Reginald Dorman-Smith, was told to buy and store between three and five thousand tractors and implements. They were all used.

'Without such an increase in tractor power,' Keith Murray the official agricultural historian of the Second World War wrote, 'it would have been impossible to carry out the task of increasing the arable areas by some six million acres. The time for ploughing and preparing the land, both in autumn and spring, was far too limited to have permitted such an expansion even if horses had been available.'

The six years of war were so fraught and busy there was no time to reorganize buildings, or the operating and lay-out of the farm, to suit engines and the huge army of tractors and ploughs assisted rather than superseded the horses. From the beginning to the end of the war the Government allowed fodder for working horses but took no interest in their breeding except for a period in 1942 when, uncertain of the war's outcome, stallion grants were resumed and rations allocated during the thirteen weeks they spent covering the mares in spring.

The end of the war in 1946 brought relief to all except the horse. 'You can pin-point the thing, you know. 1946. War ended. Things settled down. Then wham. All collapsed overnight,' reminisced a London horse-contractor to Keith Chivers in *The Shire Horse*. When people started re-equipping they went for the motor-engine and dealers, repositories, forage merchants and harness-makers went bust. In an effort to survive, the Elephant and Castle, 'the world's largest horse auction' founded in 1895, began selling second-hand cars alongside second-hand horses but eventually just sold cars.

On the farm the change was equally dramatic. Farmworkers were leaving the land but even those that remained demanded more pay for fewer hours of work. Farmers took a hard look at the productiveness of their work force, especially at the carter and ploughman who spent long hours at the beginning and end of each day baiting and caring for the horses. A motor engine could haul twenty tons per man, a horse two, and with the end of the war, there were enough machines to go round and these had improved out of all recognition. In spite of all this, for a long time, many farmers hedged their bets, keeping one horse for the occasional carting job or in case the tractor failed. Many also had the feeling that farming was not farming without a horse. John Stewart Collis in *The Worm Forgives the Plough* sums up the satisfaction which is gained from ploughing behind a horse:

> All the body is engaged, and all the mind, while the eye keeps watch on the horses and the plough, fascinated by the way the solid soil leaps up into a seeming fluid wave to fall immediately into stillness again in your wake – a green wave rising when on ley, light-brown on stubble, grey on stony ground. It falls and falls away, this little earthy breaker, until quite soon you see that a section of your field has turned colour completely, and you say to yourself – 'I've ploughed that much'.

In spite of this nostalgia the decline continued and, in the late 1950s and early 1960s, many prophesied that the heavy horse would end up a zoo animal. It did not happen, instead the tide reversed. An EEC report in 1958 drew the conclusion that increased mechanization could not meet all

138

agricultural needs especially on farms on heavy land. In places like East Anglia and the midlands a farmer with a couple of horses is able to drill his spring seed when the land is still so wet his neighbour with only a tractor could not begin for fear of getting bogged down. Similarly horses do not get stuck when hoeing, harrowing and carting early vegetable crops on low-lying ground. Heavy-horses were also rediscovered by the big brewers as being more economical than vans for local deliveries and a wonderful, spectacular advertisement for their wares. Thwaites Brewery was one of the first to re-introduce the horse and was followed by Youngs and Courage. By 1968 numbers everywhere were rising again and in 1971 the southern counties formed a Ploughing Association and within eighteen months had 263, mostly young, members. In 1979, backed by the Agricultural Industries Training Board, Joe Henson and Charles Pinney started their Carthorse Company with the aim of educating people in the handling and use of heavy horses.

A fully-trained horse today costs £1,500 and lasts 20 years whereas a tractor can cost anything between £6000 and £15,000 and soon wears out. Horse-shoes cost £10 a pair, a set of tractor tyres £200 and, at the end of the day, the knacker will pay £500 for a Shire whereas a tractor is only fit for the scrap heap. If large, six-horse hitches and American gang ploughs (developed for the prairies) are used a horse's daily ploughing rate can be increased from one acre to five or six; this does not compare too badly with a tractor's ten and, since you can plough with a horse in all weathers, at the end of the year as much work will have been done.

Heavy-horses now have new functions. Geoff Morton keeps twenty-eight on his 138 acre farm, the surplus to the five or six actually required are not there, as in former days, for sale to town dealers for draught work, but are hired out to film and television companies, rallies and even funerals. Suffolk horses are used as part of a Borstal scheme to help boys from problem homes. The boys assist in breaking the horses, walking them five or six miles a day and, during the process, they build up an emotional involvement with the trusty, reliable animal which restores their equilibrium and faith in life. Aberdeen District Council bought a pair of Clydesdales in January 1981, in preference to a thirty-five hundredweight van, for short haulage work in the city. The estimated annual saving to the council was £5000.

Many horses are going back to farms, also. A few years ago any horse on a farm would have been a relic from the old days but now they are being bought to complement tractors and in the last ten years prices have risen steeply. It seems that this is where the heavy-horse's future really lies, and there is a new breed to meet this demand. The Ardennes, imported from France, a small, stocky, dun-coloured breed with a calm temperament, quick intelligence and capable of all types of light farm work. The machinery they pull is often light as well since the dearth in this country has been overcome by imports from Poland, where, two million horses of a smaller type still work the land.

Ultimately what breed is chosen will be determined by the work and the land. Small hill farms need small, sturdy animals whereas big flat land, which can accommodate large machinery, needs large horses. But all should have well-shaped solid hooves with good quality, hard-wearing horn, short legs and hocks, a body with plenty of heart and lung room and a big belly. Draught-horses should have a well-set head and correctly positioned tail, be solid-looking with plenty of muscular development, strong bone and big flat joints. The shoulders should not slope excessively and should have a good place for the collar to enable them to pull strongly.

It was the mute strength of heavy-horses that sowed the idea of exploding the Soviet myth in George Orwell's mind. Their ingenuous willingness was symbolized in *Animal Farm* by Boxer and his maxim 'I will work harder'. To be prepared to break its heart in labour for others is typical of the heavy-horse. It would be depressing if such valour did become obsolete in the modern rush for speed, if the countryside was deprived of the presence of these passive stoics, gentle giants, towering above man their master, chomping at their bits, jingling their harness, lowering head and neck to throw their weight into the collar, lifting massive hooves to move the implements that helped create cultivated Britain.

Poultry

Swans, peacocks, pigeons, guinea-fowl, pheasants, partridges, various water-fowl and many small birds as well as hens, geese, ducks and turkeys were all habitués of the poultry yard until the end of the 18th century. In early times this varied bunch hovered on the borderline between wild and tame. Sometimes wild birds were captured and fattened in cages while, on other occasions, their eggs were collected, hatched under broody hens and the chicks became tame when reared in captivity. These, in turn, would either be fattened for food or kept for breeding and so, gradually, all the species that eventually became domesticated were acquired.

Hens were introduced to Britain by the Belgae in about 1000 BC, geese by the Romans in 55 AD and ducks by the Normans in the 11th century, while throughout the centuries the nobility kept numbers of pigeons to supply them with fresh meat in winter as well as manure. Dovecots were the prerogative of the manor and the peasants bitterly resented the plundering of their hard-won crops by these 'great devourers of corn'. Peacocks were valued both for their feathers and their meat and no feast was considered complete without one. All manner of water-fowl including swans, herons, and cranes were fattened in coops on oats and pease and were commonly eaten. Partridges and pheasants were hatched by hens and reared under nut trees in the orchard, while quails and many other small birds were caught and plumped up in cages. Guinea-fowl are thought to have come in during the 15th century, as birds referred to as Africanus were bought at Banbury market in 1407. Turkeys arrived from Mexico in the 16th century and with their introduction the stage was set for the beginning of the modern poultry yard. Then, gradually, some of the players fell out of favour.

Peacocks became purely ornamental. Swans were made royal birds and their cygnets eaten only once a year at a ceremonial dinner held at the

Vintners Hall in London. Dovecots delighted by their architecture more than by their birds. In the 20th century it would be unusual to find any birds in a poultry yard besides hens, geese, ducks and turkeys and therefore it is these birds whose history is traced in this book.

Hens

When, in about 1500 BC, the Aryans in India ceased their nomadic wanderings and began to make settlements they captured birds to supplement the food provided by their cattle and crops. The domestication of these birds began when they sometimes took more than they required for immediate consumption and kept the surplus in cages. Such birds became tame and later their numbers were added to with eggs and young birds taken from wild nests. Among these species was the Jungle Fowl, a small, lean, reddish-brown bird with a black breast and loud crow. Its terrain extended from the Himalayas down to the Philippines and from Pakistan across to Indonesia and it was naturally extremely pugnacious.

The Aryans, it is thought, soon discovered this trait and watching two Jungle Fowl fight became a pleasing diversion from the hard labours of their daily life. They did also eat the hens' eggs but these were considered a subsidiary asset. In a period when war was the normal order of events these warrior-like birds probably appealed to invading soldiers who took specimens home. By 1400 BC domestic fighting-cocks had reached China and from there were taken to Japan. The Persians introduced fowl to their country after conquering India in 537 BC. Persians believed that the 'cock, shall all the enemies of goodness overcome; his voice scatters evil' and wherever a Persian settled he took as much care to procure one as to pray and wash at sunrise. After Alexander conquered Persia the fighting-cock was taken to Greece and was called the Persian Bird. The Greeks also made it a symbol; Athena cautioned the Athenians against civil war by comparing it to the pointless combat of two fighting cocks. From Greece the cock migrated to Rome where it again signified the coming of dawn. Peter, in the Bible, denies all knowledge of Christ 'Before the cock crow'. From Rome fowl spread through Europe and eventually came to Britain with the Celtic Belgae.

The three signs that a Celtic farm was occupied were little children, dogs and cocks. The latter could be impounded if they were found in cornfields during the fortnight after sowing and later when the grain had formed. Large numbers were kept for fighting. The hens' eggs were eaten but never their bodies as, bred for fighting not fatness, they had hardly any flesh.

The Roman appetite changed this situation. They had become interested in hens with quiet temperaments, which had an abundance of flesh and laid many eggs. The Romans probably introduced their own type of hen, one whose organs had changed because of selective breeding. Organs or limbs grow in size or diminish according to their use. Darwin called this process 'correlative variability' and in hens it is most clearly seen in the relationship between the egg-organs and the comb. Laying breeds normally have large combs which develop early. The male's comb is at its largest and brightest during the breeding season, while the hen's comb grows as she comes into lay and shrinks and dulls when she stops.

The Romans did not select hens for eating which had vigour, courage and strength of bone, but those that were compact in form with broad breasts carrying quantities of meat. Pliny in his *Historia Naturalis* published in AD 77 describes a large fowl with characteristics found to this day in northern Italian hens and in the most ancient British breed, the Red Dorking. He said that 'Superiority of breed in hens is indicated by an upright comb, sometimes double black wings, ruddy visage, and an odd number of toes.'

In Saxon times each gebur had a chicken yard and was obliged to give the thane two hens at Martinmas. The Normans increased this due to include eggs at Easter and perhaps this is where the habit of giving Easter eggs originated. The number the Lord demanded varied from five to several hundred, depending on the size of the holding. They were collected by a servant and anyone who did not have them ready within twelve days was liable to a fine of twenty-one old pence. The Lord also kept his own poultry yard, situated in an inner courtyard of the castle. Five hens were allowed to each cock, a lower proportion than today when from twelve to sixteen hens are given to each male. Hens were expected to yield 115 eggs a year and 7 chicks; this is low compared to today when hens can give between 150 and 300 eggs a year and a broody sitting on 12 eggs should hatch most. Mother hens, with chicks, were often not put in a coop but tethered by one leg to a peg driven into the ground. Eggs were eaten either poached, fried or boiled and nice etiquette opined that 'one egg is gentility, two sufficient and more excess' and 'all eggs hard roasted be grosse meat'. Hens' meat was now also common fare and they were stuffed and roasted and made into dishes with picturesque names. Hindle Wakes or Hen of the Wake was a bird stuffed with fruit and spices and garnished with prunes and lemon.

Chaucer gives a vibrant picture of the mediaeval poultry yard in *The Nun's Priest's Tale*:

> She had a yard that was enclosed about
> By a stockade and a dry ditch without,
> In which she kept a cock called Chanticleer,
> In all the land for crowing he'd no peer;
> His voice was jollier than the organ blowing . . .
>
> His comb was redder than fine coral, tall
> and battlemented like a castle wall,
> His bill was black and shone as bright as jet,
> Like azure were his legs and they were jet.
> On azure toes with nails of lilywhite
> Like burnished gold his feathers, flaming bright.

Poor Chanticleer met an untimely end, the victim of a wily fox. As the hen

143

became domesticated its enemies changed. Wild fowl roosted in trees and flew strongly and their main predators were eagles and hawks but the wing-power of the mediaeval fowl was modified by domesticity, it became a terrestrial bird, and its enemies were foxes, cats, rats and weasels. To prevent their ravages John Partridge, a 17th century writer, advised 'if they eate the lungs or lights of a Foxe, the Foxes shall not eate them'.

By the 16th century there were hens of more varied colour in the poultry yard but no one knows how this came about as there is little knowledge of what caused the development of colour in plumage. Since the purpose of feathers is to act as camouflage, their colour is probably influenced by environment. Soil composition affects the shade of the legs and flesh and maybe the feathers. Food, light, dark, heat and cold are further determining factors. In the Arctic animals have no colour, partly as protection but partly because of a lack of light and all richly-coloured species live in the tropics.

Leonard Mascall writing in the 16th century recommended cocks with golden neck feathers. He considered partly black fowls and those with a tuft of feathers on the head 'reasonable good' but disliked grey or white ones. However, the shepherd's wife in Barclay's *Second Eclogue* had a grey hen:

My wife's gray hen one egge layde every day,
My wife fed her well to cause her two to lay,
But when she was fat then layde she none at all.

The essence of egg production is summed up in these three lines. Hens are born with about five or six thousand microscopic egg germs in their ovary which will not develop into finished eggs if they are fed too much food or food low in protein. Hens begin laying at about six months and the bulk of their eggs are produced in the first two seasons. Certain situations stimulate laying. Since eggs are primarily laid for reproduction their constant removal encourages further laying. The control of light also has an effect. Hens naturally begin laying in spring taking their starting signal from the lengthening daylight. This light, passing through the eye, activates the pituitary gland to secrete the hormones which start the ovary working. Hens lay at peak production when the daylight lasts fourteen hours and, in order to obtain eggs throughout the year, modern poultry keepers use electric light to persuade the small-brained hen that it is still summer.

The influence of light was known about in the 16th century as Leonard Mascall recommended hen-houses with east windows so that the inmates would wake early. He clipped his hens' wings to prevent them straying and was the first writer to describe practical incubation although Thomas More, in his *Utopia* of 1516, had already surmised that hens do not hatch eggs by sitting on them but rather 'by keepynge theym in a certayne equall heate they brynge lyfe into them and hatche theym'. Mascall advised packing the eggs in hen-droppings and bags of feathers and hatching them out 'in an oven always of temperate heat', turning them often and washing them in luke-warm water on the 18th day to moisten the shell and make it easier for the chick to break through.

Once it was advanced, the idea of incubation was developed in spite of sceptics such as Stevenson who, in 1661, said that worthless chicks were produced by those who 'set eggs in stows or ovens in winter'. Today incubation is an exact science.

Such preciseness was beyond the 17th century poultry keeper. Most of them still hatched their eggs under a broody hen, marked one side of the eggs to make sure the hen turned them, and perfumed the nest with brimstone and rosemary. The main profit of the farm came from cows and sheep, and poultry were generally a sideline, there to provide eggs and household meat. Most contemporary manuals on the subject restrict their advice to fattening methods and, to this end, hens were kept in darkened houses, fed on blood, bran, boiled root-vegetables, bread-ends soaked in beer or skimmed milk and sometimes crammed with wheat or barley-meal and milk. Some poultry were sold at market but prices were high as they were relatively scarce and considered a luxury. Poultry brought to London could only be sold at shops owned by Freemen of the City. These had been established since the middle ages at the east end of Cheapside in a part still called Poultry. After the Great Fire of London in 1666 these restrictions broke down and four other markets were set up; Leadenhall, the Stocks, Newgate and Honey Lane. Leadenhall is still a place to buy fine birds.

Cockspur Street and Cock Lane also owe their names to a connection with fowl; both were used as theatres for cock-fighting. Until it was abolished in 1849, cock-fighting was a national sport. Everyone from royalty down was an enthusiast; Henry VIII built a cockpit at Westminster and Charles II was present when two Aseel cocks from India were fought at Newmarket. The sport was justified on the grounds that the birds enjoyed it for ' 'tis their nature'. Certainly with some fowl the urge to fight remained deep in their nature. A writer called Atkinson said in the 19th century that:

> Fighting is the great bugbear of rearing Asil . . . How many times have I had a fine brood of strong healthy chicks just growing and doing well, and have returned to find all of them in a 'bloody mess' and raw heads swollen up so that they can neither see nor eat and perhaps some with beaks irretrievably damaged or torn off altogether.

The removal of their natural spurs for substitute steel ones was thought to be kind as it 'shortened the period of suffering'. Many treatises were devoted to the Fighting Cock and Gervase Markham recommended that he should

> bee of a proud and upright shape, with a small head, like unto a Spar-hawke, a quick, large eye, and a strong, big crookt beake . . . his spurs long, rough, and sharpe, a little bending and looking inward . . . For his courage, you shall observe it in his walk, by his treading, and the pride of his going, and in the pen by his oft crowing.

One of the most extraordinary aspects of cock-fighting was the regular fee, or tax, schoolmasters received for allowing boys to bring Game cocks to school 'to amuse themselves the whole morning watching their encounters'

145

on Shrove Tuesday. In 1790 the income of a schoolmaster in Applecross, Ross-shire, was made up of 'salary fees and cock-fighting dues' but by the 19th century the custom was largely abandoned with the practice of cock-fighting. It lingered on longest in the north-west of England and is still legal in France.

The breeding of domestic fowl was a separate occupation and Markham had as much advice to offer on it saying that cocks were to be kept loose and that it 'is a fowle of all other birds the most manliest; stately and maiesticall'. Hens he said should be as,

> perfectly bred as your Cocke . . . her head would be smal, her eye very cheerfull and her Crowne armed with a double Cofpel or Crownet; her body would be large for so shee will cover her Broode the better, and the feathers on her brest would be long and downie, for that is most comfort to the Chickens . . . Lastly looke that she be a painfulle layer, a willing Sitter; and above all things loving and kinde to her Broode.

In spite of all Markham's instruction hens were still a motley bunch distinguished only by shape or colour. They all went under the generic name of Barn-door fowls or, the less attractive, 'dung-hill fowls'. Cottagers lived in close proximity to their poultry, brought them in to spend the night in front of the fire and 'in consequence', wrote Daniel in his *Rural Sports* of 1813, 'farmers whose poultry are in the night time confined in places without a fire, obtain no eggs, while the poor people have them in abundance'. Daniel also states that in Scotland 'Poultry are reared in vast quantities and several cartloads of the eggs of dung-hill fowls were annually collected by egglers who would sell them in Berwick, for the London market'. These merchants also bought hens and were selective in their spending; they gave more money for the better type of fowl and it is due to their choosiness that definite breeds began. In Scotland a cuckoo-patterned grey hen emerged, now called a Scots Grey. They are great layers of white-shelled eggs and became the most widely-spread barnyard fowl in the lowlands and central Scotland. A similar hen, the Scots Dumpy, named after its short legs and rolling gait, was bred in the Highlands.

The London meat market preferred 'the kind' of hen 'which has white feathers and legs and which I had heard say carried flesh of a much finer grain than the larger sorts with other coloured legs and feathers'. White hens to fit this taste were raised all over the Weald of Sussex but reached their greatest perfection round Dorking and so gained the name of Darking or Dorking Fowl. The breed was in all likelihood based on the fat, red fowl described by Pliny and Columnella and brought here by the Romans but, by the 18th century, they had become white and had 'the credit of supplying the market with the finest specimens both for appearance and the table'.

In other parts of southern England other fowl were given specialized attention to fit them for refined London tables. In areas outside the Weald in Sussex, in Surrey, and to an extent in Lincolnshire a fowl similar to the Dorking was bred, but with four toes not five, and called the 'Old Sussex or

Kent fowl'. In Devon and Cornwall they bred the Cornish Game fowl. This was later named the Indian Game fowl and was probably a relic of the hen brought by the Belgae. They had a yellow, fuller-flavoured, richer flesh 'hard in texture, close in grain' and laid a small number of dark-brown eggs. This large boned, slow-maturing hen was marketed around Launceston, Callington, Liskeard and Bodmin and was taken to America by the early pilgrims. The true Cornish Game hen now no longer exists because of crossing, primarily with Indian Game hens, which is how the name changed.

In the north, when the Romans departed, they left some of the red Dorking-type hens they had used to feed their garrisons in Cumberland and along the border of Scotland, and an enthusiasm for breeding hens which was much later fostered by the effects of the Industrial Revolution. Country people, forced to work in towns, missed their livestock, but the space between their back-to-back houses was too small for a cow; however, it was large enough for a few chickens. Perhaps as a way of enlivening the occupation of keeping poultry they began specializing in breeding fancy chickens, selecting those with elaborate combs and patterned plumage. Then they began staging local competitions in inns and public houses. The competitions were staged in a unique way; Edward Brown in his book *Races of Domestic Poultry* (1906) describes how one for Spangled Hamburghs took place.

> It was the custom for the exhibitors to take their birds under their arms to a room and show them on a table, each exhibitor arguing for the good points of his birds, whilst his rivals pointed out their defects. The judge stood at one side of the table and heard all that was said *pro* and *con* for each bird, of which two were on the table at one time. The worst of these was taken away, the one left having to face the next competitor, and so on. That left at the conclusion of the contest was declared the winner. This system, which would be utterly impractical nowadays, doubtless had much to do with the perfection to which the breed was brought, for each breeder learnt just in what way his bird was deficient, and he could seek to remedy the defect.

Many of the competitors were textile workers who did business on the Continent and, on trading trips, they saw and brought home new varieties of hen. In this way the Poland, a shining black fowl with a flurry of white crest feathers, arrived in Britain. The Poland probably originated in Asia and spread through Russia into Europe. Selective breeding has increased the crest, which nature gave it as protection against cold, to a magnificent size. But such is its weight that the bird has to be kept under cover as, were it to get wet, its head would become top-heavy and it would probably catch cold from the sodden feathers.

Birds from Holland contributed to the make-up of the Hamburgh but most of its blood came from Yorkshire and Lancashire hens, described by the great 19th century authority on poultry, Bonnington Moubray, as 'small-sized, short in leg and plump in the make. The colour of the genuine kind, invariably pure white in the whole lappel of the neck; the body white, thickly spotted with bright black, sometimes running into a grizzle, with one or more black bars at the extremity of the tail. They are chiefly esteemed as very constant layers.' Another northern breed was the Derbyshire Redcap, named after its rosette of comb wattles that resembled a cloth cap. This small, plump, rich brown and black hen was bred on hill farms in Derbyshire and Yorkshire. Besides these breeds many other varieties emerged that were called, variously, Moonies, Bolton Greys, Bays, Creoles, Moss and Pheasants.

By the beginning of the 19th century there were three distinct divisions of poultry. The south produced fat fowl for the luxury London market, the north their fancy varieties and, everywhere else, there were dung-hill fowls picking a living round the farmyard. A radical change came about when the hens ships' captains had carried on board to provide eggs and meat during the voyage and which they sold on arrival at port began to excite interest amongst zoologists and some breeders. Many of the fowl were bought in the East and their strange looks stimulated zoologists, fired by the discoveries of Darwin, and poultry breeders infected with an enthusiasm for the new methods of breeding practised on other farm animals, to try their theories on these fowl.

In 1815 Bonnington Moubray formalized the new ideas on poultry breeding in his book *A Practical Treatise on Breeding, Rearing and Fattening all Kinds of Domestic Poultry*. It specified twelve breeds and laid down rules for poultry-keeping which revolutionized the industry and gave poultry a foothold on the orthodox farming map. Poultry were made fashionable when, in 1845, Queen Victoria was presented with some Cochin fowl acquired off a ship from Shanghai. Queen Victoria exhibited them the following year at the Royal Dublin Society Show. They caused a sensation with their majestic form, their profuse golden plumage, heavily-feathered legs and resonant voices (compared to the roar of a lion). The Cochins, furthermore, had the attribute of laying brown eggs in winter and this, coupled with their looks, caused them to be 'vaunted to the skies as the best, the most prolific and handsomest bird that was ever seen'. A craze for chickens gripped the country. Queen Victoria's lead was followed by other members of the aristocracy. Poultry-shows were 'honoured by the patronage of Lords and Ladies . . . Earls, Marquises and Dukes' all of whom 'shared in the glory of the exhibition and on the day of sale competed for the possession of the meritorious birds'.

The demand for specimen birds encouraged the breeding of ones with fancy points and Edward Brown says 'extremes were sought for and variations stimulated, whilst the world was searched for rare or novel types, resulting in the introduction of alien races, many of which have, deservedly, by their valuable qualities, attained a greater amount of popularity than some of the native breeds.'

The Far East yielded the Brahma, a silvery white and black, huge, tea-cosy shaped bird which emerged from the Brahmaputra district of India.

The Langshan, a round-bodied, brilliantly-plumaged, handsome black, white or blue bird came from the district north of the Yang-tsze-kiang River in China. The Malay, a red, narrow, fierce-looking bird 'cruel by nature' arrived from Malaysia and the Black Sumatra, a pheasant-shaped, fighting bird similar to the Malay from the nearby island of Sumatra.

Two light-bodied, sparse-fleshed, long-legged, quick-maturing, heavy-laying birds came from Italy. The Leghorn arrived in Britain, via America, from the Leghorn district in Tuscany; and the Ancona, a nervous bird but 'one of the best layers' came from the area round the port of Ancona in eastern Italy.

France produced birds looking 'somewhat mediocre to the eye of the fancier' but which made excellent eating as with 'the French the economic qualities ever stood first'. Practically every village in France had its own variety of hen, several of which were introduced to Britain, but the only one to become popular was the Houdan or Normandy fowl. This general-utility bird, producing both eggs and a good carcase which was fattened for the Paris market, is an extraordinary-looking, speckled bird with five toes, a triple comb and a 'Paris hat' of extravagant head feathers.

Spain yielded three light-bodied, egg-producing birds. One was the Black Spanish, an ancient, black-feathered, red-combed breed whose flesh is 'scanty and dry' but whose numerous eggs are large and white-shelled. A second breed was the Minorca which was probably first introduced to England by Spanish and French prisoners-of-war interned at Tiverton in Devon in 1780. More importations were made by soldiers returning from the Peninsular War of 1813–14 of this rather delicate black fowl with a large, drooping, vivid scarlet comb and white ear lobes. The Minorca looks similar to the Spanish but is rounder and lower in the leg and is 'par excellence an egg producer'. The third breed to arrive from Spain was the Andalusian; a finely marked, greyish-white bird, another fine egg-producer.

The Campine was the only Belgian breed to be exported to England in spite of the numerous, practical hens bred by these industrious people to whom 'the aesthetic appeals . . . to a very limited degree'. An ancient type of hen with gold and silver plumage, the Campine is found all over the Low Countries but takes its name from a dry, sandy plain lying between Antwerp and Hassalt. Its egg-laying average can exceed 200 a year and its flesh at a young age is 'wonderfully delicate' and made into 'poulets de lait'.

Chicken meat has never been very popular in Germany and their contribution to British poultry is confined to one breed, the Lakenfelder. This is an egg-laying fowl from the district of Hamburgh, named after its body plumage which resembles a white 'laken' overlaid by a black 'feld' of neck and tail feathers.

All these birds caused a flurry of colour and excitement in the exhibition halls. Hot debates raged over the niceties between combs that were single, rose, pea or horned shape, flesh that was white, yellow or grey, legs of white, pinky-white, dark or slate-blue colour, eggs that were white or tinted and feather markings which could be barred, laced, spangled, cuckoo, pencilled or chain-mail. And bantams, specially-bred, diminutive replicas of the larger breeds became fashionable.

Once the first enthusiasm abated and the economic properties of the new hens were more clearly examined many were found wanting. Under a different climate and terrain some of the excellent characteristics they exhibited in their country of origin changed, or were changed by breeders concentrating on producing handsome birds for the show bench. Monstrous combs, it transpired, flopped over, blinding the bird in one eye, pulling the head sideways and, by the pressure their weight put on the brain, eventually crazing and killing the bird. Cochins became worthless when it was seen that the eggs they laid were few and small and the flesh of older birds tough and inedible. The craze for chickens passed. Many of the new breeds vanished almost as soon as they arrived and those that survived became classified into four types. (1) Laying breeds which were usually quick growers of medium-size and active habit who laid white-shelled eggs. These hens have little maternal instinct and are poor sitters, as their broodiness has been bred out of them over years of being encouraged to lay rather than hatch eggs. Into this category fall the Ancona, Campine, Minorca, Leghorn, Redcap, Hamburgh, Scots Grey, Andalusian, Welsummer and Houdan. (2) Table breeds, mostly large hens of sedentary habit which grow and fatten readily. They are good mothers and enjoy nestling over a brood of eggs. Breeds in this category include the Dorking, Indian Game, La Bresse, Faverolle and Creve Coeur. (3) General Purpose breeds, large birds of heavy bone who grow slowly but have good flesh and lay coloured eggs. These hens are good sitters and mothers and include among their numbers the Orpington, Plymouth Rock, Wyandotte, Langshan, Rhode Island Red, Light Sussex and Maran. (4) Fancy breeds, of which there are too many to mention.

In the late 19th century breeders went to America in search of poultry strains whose good points had not been spoiled by the show ring. Here the Old English Fowl imported by early settlers and the new Asian and Continental breeds had been selectively bred for different points. The Americans preferred yellow flesh to white and wanted hardy, dual-purpose, large-bodied birds with clean-feathered legs and active habits. They achieved these ends by crossing the different varieties. The stylish, black and white barred Plymouth Rock came out of a melting pot of Cochin, Dorking and Malay; the Wyandotte a quick maturing hen, which laid eggs in winter, emerged from a mixture of Brahma, Cochin and Hamburgh; while on the shores of Narragansett Bay in New England the Rhode Island Red, a strong and vigorous hen, came from the crossing of Cochin, Malay, Leghorn and Wyandotte. In due course all these practical breeds found their way into Britain.

Shows also became more practical in outlook and began to include classes for 'exhibition-utility' hens and later ones for 'utility' only. Fowl for the table had to have all the flesh on the breast and not on the legs. Furthermore the flesh was to be soft and white, not close or yellow, and to cover a fine-boned, lightly-feathered frame which matured early as such bodies were produced with less energy and feed. Egg-laying trials were instigated. Now only those breeds laying up to 300 eggs, weighing at least 2 ozs each, were considered worth keeping. The White Leghorn, Light Sussex, White Wyandotte, Rhode Island Red and the super-laying Australoup from Australia and Barnevelder from Holland came top in these laying trials and until the 1940s were the most popular breeds.

Until the First World War most hens still scavenged a living from the farmyard and were kept mainly to provide the family with eggs and meat. But during the war egg prices rose and farmers with an eye to profit began to take poultry more seriously. Then, after the war, with prices still high, many would-be-farmers with little capital went into poultry.

Thousands of small poultry farms sprang into being and many established farmers also built units. Few of these ventures were successful. The depression of the 1920s and 30s bankrupted some, while others were destroyed by the foibles of the birds. After centuries of leading a free-ranging life poultry objected to being herded into close flocks, in cramped pens, and often fed inadequate food. Many died and still more were killed by disease which became prevalent when, to meet increased demand, hens were indiscriminately bred and their stamina reduced. By the late 1930s many of these problems had been resolved but the Second World War brought new ones; primarily that of a scarcity of feed. This swung poultry-keeping away from the specialist and into the sphere of the general farmer, who could grow the necessary grain. The Second World War also produced a resurgence of back-yard poultry fed on household scraps.

After the war, a similar situation arose to that after the First World War. New recruits rushed into the business, but learning from the mistakes of their predecessors they had more success and the production of eggs and meat became increasingly efficient. In 1948 roughly 95% of laying fowl were kept under range conditions; usually in 'fold units', or small portable houses with wire runs attached. These houses were moved each day over grassland and the hens manured the ground in much the same way as hurdled sheep. This method required a certain amount of labour and, as the 1950s began, workers became expensive and scarce. As a result hens were increasingly herded into large straw-covered yards or large houses. With time poultry keepers devised neater, more manageable and specialized methods of production. Then that bogey of animal life, science, entered the scene. Broody hens were abolished and chickens were mass-hatched in huge, electric incubators. They were bred to formulae, each poultry empire concocting its own secret recipé of bird suited to its system and the hens were given trade rather than breed names. Genetics meant that sexing no longer had to be done by examination, because it was discovered that if you cross a dark-coloured cock with a white mother all the cockerel chicks will be white and all the pullets dark. Specialists in each department of poultry production arose. Those who incubated the eggs sold the chicks as day-olds to people who reared them to about six or eight weeks old and then sold them again to battery egg or meat producers.

Those destined to lay eggs now live in benign concentration camps for the whole of their short lives. At twenty-one weeks old they are stuffed into small wire cages no larger than 450 cm square, set in tiered rows, under fluorescent lighting, in a heat and humidity-controlled building. The hens are mechanically fed a constant supply of compound feed and water and when they lay an egg it rolls into a gutter outside the cage for automated collection. In these units one man can tend 10,000 birds and during the 60 weeks the hens spend in their cages, they are expected to produce between 230 and 300 eggs. The hens emerge from their battery period deformed and featherless. More than 90% of the eggs consumed in Britain in 1981 were produced under these conditions. Public outcry at its iniquities and a resistance to buying battery eggs are forcing some poultry producers to look at alternatives. But most prefer to argue that the hens are happier and better-cared-for under the system since the total capital cost of replacing all laying batteries with deep litter systems would be £300–700 million. Unless the EEC revises its rules beyond merely enlarging cage sizes slightly, and stating that hens have to be inspected daily by a vet, for the time being batteries are here to stay.

The meat bird has a different fate. A few are reared as roasters (3–5 month old chickens weighing at least 5 lbs) or capons (6–7 month old castrated males of weights over 7 lbs) but most become broilers, processed at eight weeks old for frying, roasting or broiling. The hens are hybrids of the flesh-producing breeds, preferably white-feathered for ease of dressing, and with pale carcases which is what the consumers like. The birds are vaccinated through their feed against diseases such as *coccidiosis*, which damages the intestinal lining and impairs growth and food conversion, and are kept twenty to the square metre in small wire cages set in tiers in semi-dark houses. Just enough light is allowed to enable the bird to see to move about, eat and drink, and the temperature is carefully monitored as cold birds eat more and convert the food less efficiently into meat. Feeling hot, cramped and bored the hens become aggressive and peck each other, often to death. As a result most are now debeaked which makes it difficult for the hen to grasp feathers; the process involves cutting away more than half the upper beak and shortening the lower. Because they spend their time crouched on wire mesh their claws curl and their legs become deformed. Once they are sufficiently fat the birds are stuffed into crates, wings and legs becoming bashed and broken in the flurry, and shipped to factories to meet a demeaning end. There they are hung by their legs from a moving pulley which carries them relentlessly through an electric stunning plant which, if it fails, means the bird dies from being immersed in the boiling water tanks designed to loosen the feathers for plucking. Every week six million chickens die in this way in Britain. In 1980, the quantity of poultry meat that was sold, mainly as chicken portions or cooked products, amounted to 418,000 tonnes. Poultry has overtaken lamb and pork as the most popular meat and, by 1986, is expected to have exceeded beef.

The consumption of eggs has also risen by 87% in recent years and over half the poultry in Britain are kept for egg production. To meet these demands some form of intensive farming is obviously necessary. But some alleviation of the living conditions of battery hens is necessary and a start would be for more poultry farmers to try the new deep-litter systems and aviaries where the birds are kept free and stocked at lower densities on tiered platforms.

Some poultry still have the freedom of the farmyard and others are kept as of old in back-yards in Lancashire and Yorkshire. Here proud fanciers continue to show hens of paint-box variety and the nobility still lend their support.

Geese

The goose in its wild form is a denizen of the lakes and swamps of Europe and North Africa and, long before it was tamed, it was hunted. Domestication did not take place until man became settled, as a nomadic life does not allow for birds sitting for thirty days to hatch young which then take more time to rear. Once stable communities were established, those by lake shores in Central Asia probably took young geese from their nests, clipped their wings and tried to rear them. This wild goose was most likely a Greylag (*Anser ferus*) as they are common all over Europe. The assumption that it was not the equally prevalent Bean, Whitefoot or Pinkfoot is based on the close resemblance between the greylag and the domestic goose. When kept together they will associate and mate whereas the other three breeds stick strictly to their own kind.

Geese were already domesticated by 4000 BC and, in other countries, different varieties were taken into captivity. Egyptian monuments of the time show both Greylag and Egyptian geese. The Romans regarded geese as sacred to Juno, the protectress of marriage and women, and, as is well known, their cackling saved the Capitol from a surprise attack by the Gauls during a siege in 365 BC. Geese were also eaten by the Romans. Pliny is wonderfully explicit on the subject saying:

> our people esteem the goose chiefly on account of the excellence of the liver, which attains a very large size when the bird is crammed. When the liver is thoroughly soaked in honey and milk, it becomes specially large. It is a moot question who first made such an excellent discovery ... This bird, wonderful to relate, comes all the way from the Morini to Rome on its own feet: the weary geese are placed in front, and those following by a natural pressure urge them on.

The Romans preferred white geese and selectively bred them to this end. The result was a rapid-growing, prolific-laying but quite small bird and it went with the Romans when they conquered and colonized central and eastern Europe. These geese are still found distributed throughout Southern Germany, Austria, Hungary and the Balkans. Long before this the habit of keeping domestic geese had become established in Britain. Some say that the Belgae, who introduced hens, also brought in geese. Domestic geese were obviously common in Celtic times as, under their laws, anyone finding a goose in his corn was empowered to take a stick 'as long as from his elbow to the end of his little finger and as thick as he may will' and kill the goose.

The Romans may have introduced their own white geese as Palladius mentions this colour being preferred to skewbald or brown. They certainly introduced more method into the keeping of geese as Palladius also advises that goslings should be hatched under hens sitting on nests of nettles and afterwards fed in-doors for ten days. One gander was allowed three geese which is about the same ratio as today.

Geese are scarcely mentioned in Saxon laws as they were not important members of the farmyard. They were not vital to the Normans either, although the lord's farm did sometimes have a goose-house and the flocks were driven out to graze by a goose-herd. A picture in the mediaeval *Luttrell Psalter* shows a goose-herd shooing a crow away from five goslings and the author of an anonymous *Hosebonderie* says that a dairywoman should get five goslings a year from each goose and allows five geese to a gander.

During the middle ages geese became essential because of their feathers. Arrows from the long bow, that weapon which defeated the French at Crécy and Agincourt, were steered on their deadly accurate course by goose feathers. Goose wings were used as dusters for ledges and velvets. Their quills made teats for bottle-fed lambs, fishing-floats and valves, or indicators, whereby brewers could judge the ferment in a bottle. They also held the soft bristles of artists' brushes and were made into pens. The quills were carefully prepared; first the membrane on the stem was removed by being heated in hot water or sand, next the feathers were cleaned with a cloth or dogfish skin, and finally the nib was shaped with a 'penknife'.

Goose-grease (the yellow fat which drips from a cooking bird), was used in all manner of ways; for cooking, for filling sandwiches (mixed with onions, parsley and other ingredients), to soften leather, as a poultice, a cure for gout, and an all-purpose skin-cream, when it was mixed with yellow broom and gorse flowers. The grease was also rubbed into chapped skin, cows' udders, cracked horses' hooves and sore dog pads and the beaks, hooves and trotters of animals were polished with the warmed emollient to make them shine.

Goose down, with the increasing attention paid to comfort in the middle ages, was used to fill pillows and feather beds.

Flocks were controlled by goose-herds, aided by a rattle made of paper or parchment and 'the noise whereof cometh no sooner to their ears than they fall to gaggling and haste to go with him', and they were mainly found in the goose's natural home, the eastern, low-lying marshy parts of the country. At Scoter in Lincolnshire an unenclosed piece of ground called the 'car' was traditionally reserved for geese and owners of geese everywhere were admonished to clip or pull their wing feathers to prevent them straying. Geese, generally, were fattened and eaten at two times of year. The elder geese, plump from the gleanings of the stable fields, were eaten at Michaelmas. Tusser celebrates this event in his poem on *The Farmer's Feast Days*.

> For all this good feasting yet art thou not loose
> Till thou give the Ploughman in harvest his goose.
> Though goose go in stubble, yet pause not for that,
> Let goose have a goose be shee leane be shee fat.

A full-grown goose has, according to Cobbett, a 'solidity in it; but it is hard, as well as solid; and in place of being rich, it is strong', and goes well with highly-flavoured stuffings like sage-and-onion. 'Green' geese, were eaten at Whitsun, after being fed for a month in pens or dark houses on ground malt, oats, milk and chopped carrots. Cobbett thought them 'tasteless squabs'.

The 16th century writer, John Worlidge, said that darkness was conducive to fattening and described how the Jews wrapped their geese in linen aprons, stopped their ears with peas to prevent them being distracted, and fed them three times a day with pellets of ground malt and steeped barley. The French, used to immobilize geese, by nailing their webs to the floor, and mechanically crammed them with food in order to swell their livers to make pâté-de-fois-gras.

Most geese in the 16th century were kept for their feathers and many must have reached their full life-span of about twenty years. Those birds sometimes had their feathers shorn but mostly they were plucked, five times during the summer, at six-week intervals. They were said to suffer no hardship but a 16th century scribe does describe them as 'looking piteously melancholy' by September. These geese were a motley bunch, of all colours and no particular breed and much resembled their wild ancestors.

By the 18th century as many geese were bred for their meat as their feathers. Droves were driven to the London markets and large flocks encumbered the roads out of East Anglia. Large numbers also went to the Nottingham Goose Fair. At its height this fair lasted twenty-one days and 20,000 geese passed through its portals. The geese travelled on foot and fed in stubble fields along the way. Their feet were given protection from the wear of the road by being made to walk three times through a mixture of tar, sawdust and sand. Sometimes felt pads were attached to their feet but never, as is popularly supposed, iron shoes.

This goose-orientated life came to an end when the fens were drained and enclosed. Before this, according to Lord Ernle, 'Whenever the rains fell' the fen rivers 'rose above their banks, and, especially if the wind was blowing from the east or south, flooded the country for miles around' making the whole area 'a wilderness of bogs, pools, and reed-shoals – a vast morass, from which here and there, emerged a few islands of solid earth.' Here the 'amphibious population' travelled in punts, walked on stilts and lived mainly by fishing, cutting willows, wild-fowling and keeping geese.

Moves had been made towards the reclamation of the fens from the 16th century and these were accelerated throughout the succeeding centuries by landowners longing for space on which to fatten more profitable beef cattle and sheep. But to the fenmen 'Catching pike and plucking geese were more attractive than feeding bullocks or shearing sheep' and they rebelled, attacked the new flood-gates and sluices, levelled the enclosures and broke down the embankments. In the 18th century, a fresh attempt was made on this 'wild country', that 'nurses up a race of people as wild as the fen' whose geese were the only thrifty animals, which eventually succeeded.

In other parts of the country protests against enclosures, which removed common rights of grazing, were no less vociferous:

'Tis bad enough in man or woman
To steal a goose from off a common;
But surely he's without excuse
Who steals a common from the goose.

Nevertheless geese were soon relegated to much the position they hold today, and were kept only in small units by cottagers and farmers. Cobbett wrote in the 19th century that geese 'can be kept to advantage only where there are green commons, and there they are easily kept, live to a very great age, and are amongst the hardiest animals in the world. If kept well, a goose will lay a hundred eggs in a year.' He used to fatten a flock at his home in Kensington.

In spite of these set-backs the geese began to improve in quality and specific breeds appeared. The Earl of Derby bought some geese, for his menagerie at Knowsley, from the Haute Garonne in the south of France where there were more geese than anywhere else in Europe. They were monstrous, grey birds weighing up to twenty-eight pounds with deep bodies enveloped in heavy-hanging folds of flesh which almost reached the ground. These geese were carefully bred and became famous for their many, large eggs and huge frames which 'when fattened lay on a large amount of flesh'. But their flabby paunches were against them and they only became successful when crossed with a firmer-meated breed from Germany, the Emden (formerly Bremen), a white goose of proud stature and stern eye which did not lay as many eggs as the Toulouse but whose frame fattened faster and on less food.

Another breed which arrived in the 19th century was the Chinese goose. This is now the third most popular breed in Britain but was the most favoured one in China, Siberia and much of India. Domesticated since ancient times, the first ones to come to Britain are said to have been hatched on a ship travelling from China in 1848. This goose has never been used commercially; they are smaller than the Toulouse or Emden and, although they lay a good number of eggs, have fine flesh and are hardy and make good mothers, they tend to be hysterical and quarrelsome and have harsh voices. The Chinese is very elegant-looking. It can be either white or browny-grey and has a long neck, pretty patterning and curious knobs on its nose.

Many other breeds were also imported during the 19th century, but none really became established outside zoos and water-fowl parks. Even a Russian fighting goose was imported. In Petersburg, according to Moubray 'they have no cock-pits but they have a goose-pit, where, in spring, they fight ganders, trained to the sport, and so peck at each other's shoulders till they draw blood'.

The Old English goose survives in breeds such as the Pilgrim, a medium-sized bird with the individual attribute that the ganders are always white and the geese are always grey; and the Brecon Buff, a hardy, meaty bird based on the traditional fawn-coloured geese of central Wales. But most people today keep cross-breds and the most popular is the Emden × Toulouse. In 1938 there were about 608,715 geese in the country, this figure rose during the Second World War to 937,955 but in 1970 was down to 456,531 and in 1980 was only about 100,000.

By far the greatest number are kept by smallholders or by upland farmers who have a few as a sideline. These small flocks are allowed to range at will over the fields and conjure up visions of the far north and their wild relations.

Ducks

The wild mallard is the progenitor of all domestic duck and it was very likely a Chinese who first took one into captivity. From China the habit spread, probably along the Silk Route, to Rome. By the 1st century AD, according to the Roman historian Columnella, ducks were common and he wrote that

> when anyone is desirous of establishing a duckery, it is a very old mode
> to collect the eggs of teal, mallard, etc., and to place them under
> common hens; for the young thus hatched and reared cast off their wild
> tempers, and undoubtedly breed when confined in menageries.

Ducks became larger and clumsier with domestication. It is inevitable, says Edward Brown in *Races of Domestic Poultry* that 'The beautiful carriage of the wild mallard and his mate' changes 'to the easy, well-to-do, comfortable deportment of a small Rouen, for they at each reproduction become much larger'.

Wild duck were prevalent all over the watery parts of Britain but there is no mention that any were kept in captivity until near the end of the 13th century. During Edward I's reign 288 domestic ducks are suddenly mentioned as part of the tithe paid by tenants of two manors on the Berkeley Estate. In 1305 the Council of Merton College also demanded a tithe of ducks. Ten out of their eleven manors kept ducks but, since the average tithe demanded was nine, they cannot have been plentiful.

These ducks were probably introduced from France by the Normans.

157

Alexander Neckham, who spent part of his life in Paris, includes ducks in his list of domesticated fowl, compiled in 1190, but as no other English writer of the time mentions ducks it is assumed he saw them in France. At any rate, Gaul in Normandy became famous for its ducks. The Normandy or Rouen duck closely resembles the wild mallard and has a dark, gamey flavour. It also has a massive frame, weighing ten to eleven pounds, and a large, broad head carried on a long, graceful neck. Its excellent taste is attributed not only to its breeding but to the fact that it is traditionally killed by strangulation, or smothering, to ensure that the flesh retains all its blood and flavour. A method, according to Elizabeth David in *French Provincial Cooking*, 'which would not be tolerated in this country, where we treat our animals with more consideration than we do our fellow men'.

Throughout the middle ages and 16th and 17th centuries, ducks gradually established themselves in Britain as permanent members of the poultry fraternity, but they were never highly regarded. Rather, they were denigrated as 'grosse, greedye and filthie feeders'. John Worlidge in his *Prognostiks* published in 1669 said you could tell whether it was going to be a hard winter by the colour of the duck's breast bone; 'If the brest bone of a Duck be red, it signifies a long winter, if white the contrary.'

The duck's eating habits had not risen in anyone's estimation by the 18th century as Ellis remarks that 'A good parcel of ducks will do great service in a turnip field where they are seized by the black caterpillar'. They also did great service in the Vale of Aylesbury as, about this time, this boggy region, split by rivers, started to become famous for its white ducks. The Aylesbury (formerly the White English Duck) is a boat-shaped bird with a pinky-coloured bill and feet. It matures early, has light bones padded by creamy-white flesh which is especially thick on the breast. The ducks are prolific layers and their ducklings are traditionally eaten in spring, and at Whitsun, stuffed with a delicate sage-and-onion mixture and accompanied by young peas.

The Aylesburys were mainly fattened for the London market and presumably were fed special food for people were still suspicious of the duck's eating habits. Cobbett in his *Cottage Economy* published in 1822, gives precise instructions on the raising of ducks;

> When young, they should be fed upon barley-meal, or curds, and kept in a warm place in the night-time, and not let out early in the morning. They should, if possible, be kept from water to swim in. It always does them harm, and, if intended to be sold to be killed young, they should never go near ponds, ditches, or streams. When you come to fat ducks, you must take care that they get at no filth whatever . . . I buy a troop when they are young, and put them in a pen, and feed them upon oats, cabbages, lettuces, and water, and have the place kept very clean. My ducks are, in consequence of this, a great deal more fine and delicate than any others that I know anything of.

By the end of the 18th century ducks were kept in ever-increasing numbers especially round London and loads were sent up twice a week from Peterborough and St Ives.

In other parts of the country, until the middle of the 19th century, the typical farmyard had a motley-coloured collection of ducks made up of the Rouen's mallard browns and greens and the English's white. They quacked round the farm pond and ranged over the yard and fields picking a living of whatever good or 'nauseous' articles of food they could find. In spring the ducks settled themselves over a clutch of ten to thirteen eggs for twenty-eight days and maintained a maternal vigilance over the flotilla of fluffy, yellow ducklings that emerged. Each relinquished her duties when at the onset of adolescence, the duckling's coat coarsened into adult feathers, barbed together into a shining, water-shedding carapace. At the end of the 19th century some people began to bring in exotic breeds of duck from abroad.

In 1872 white ducks with a 'canary tinge' of yellow in the plumage were imported from Peking. The birds were upright in stature and so large that an American seeing them for the first time in China mistook them for geese. In China ducks have always been loved and great expertise goes into their rearing and cooking; Peking Duck is considered, by gastronomes, to be a dish of the highest merit. In Britain the Peking Duck was first used to infuse new vigour into the Aylesbury, which had deteriorated through inbreeding, and then it was bred in its own right. But even though the ducks are hardy, excellent foragers and layers they never became commercially popular as they carry less flesh and mature some two to three weeks later than the Aylesbury. In the USA, however, these deficiencies were overlooked in view of the Peking's tolerance of extremes of temperature.

During the 1890s Ships' Captains brought in an elegant, slim-figured 'penguin-carriaged', agile Runner duck from India. This bird is a good forager and, if given the range, will find all its own food. It has an 'indifference to water and great laying power', producing as many as 200 hen-sized eggs a year until it is six or seven years old, and its scant flesh is 'fine in quality, juicy and well flavoured'. Indian Runner Ducks, as they came to be called, have never been reared commercially and are really considered fanciers' birds. There are several varieties; the Trout is tawny; another the Black East Indian is ravishingly handsome with a green sheen to its jet-black feathers. The Cayuga duck also has a green shot-satin sheen on its metallic black feathers and is a broad, meaty bird. It is said to have come from Lake Cayuga in New York State but was probably first found in South America and gained its colour from a cross with a Black East Indian.

The Muscovy definitely came from South America and is the only domestic duck not related to a mallard. The Muscovy was first described in about 1670 by a Frenchman called Cauis. He called it 'a wild Brazilian duck of the bigness of a goose'. The name could have come from a 16th century trading company, the 'Muscovite Company', as 'geographical knowledge at that time was very indefinite, and names were accorded to novel importations which were, to say the least, misleading'. Young Muscovites make excellent eating and any one is decorative with its smart black and white feathers and bright red, knobbly-fleshed face.

Ducks from Russia, Sweden and Belgium along with some crested ones from Bali and Holland were all imported during the late 19th century but none became commercial and they are only preserved by waterfowl enthusiasts. Still, their numbers are growing.

Most commercial ducks are found in Norfolk although some are kept in Aberdeen and Dumfries. There are about two million of these ducks but even so they only constitute 1·1% of all poultry as ducks, although less prone to disease than hens, eat more food and people will not buy their eggs as they still have an innate suspicion that they are unclean. Ducks, like hens, have been divided into categories; the table breeds include the Aylesbury; the laying, the Indian Runners; and the dual-purpose, the brown Kharki-Campbell. But most commercial ducks are now hybrids and they are kept in houses and specially fed. Some ducks show selectivity in mating but most are polygamous and six females are satisfied by one male. Their eggs are hatched in incubators and the ducklings reared under warm lamps. At six to seven weeks of age the sexes are separated and certain females held for breeding. Their feathers and voices give the clue as to whether they are male or female; the duck's quack is loud and harsh and the drake's soft and throaty.

Ducks are fussier than chickens and will not tolerate or develop under battery conditions. As a result they are allowed houses with litter on solid floors and more space (about five or six square feet), natural daylight and air and sometimes an outside yard. Ducks destined for the supermarket are killed at seven to nine weeks old and then neatly packaged and labelled Norfolk duckling.

Many other ducks just waddle about on farms as of old. A melée of breeds and colours, they give comfortable quacks and squawks as they unhurriedly pursue their daily business around field, barn and pond.

Turkey

The original turkey was an enormous bird, with black bustles of plumage, which roosted in the trees of Central and South America. It is unknown when they were first domesticated but, by the time the Spaniards arrived in the 16th century, they were commonly kept. In his *Conquest of Mexico* (1872) William Prescott says that

the spaniards saw immense numbers of turkeys in the domesticated state on their arrival in Mexico, where they were more common than any other poultry. They were found wild, not only in New Spain, but all along the continent, in the less frequented places from the North-Western territory of the United States to Panama. The wild turkey is larger, more beautiful, and in every way an incomparably finer bird than the tame.

In North America bronze-coloured turkeys were 'denizens of the woods and forests'. Early settlers remarked on their presence, as the pioneers moved west, they found them domesticated by the indigenous Indians, and once they became established they also took the birds into captivity. Under domestication the turkey decreased in size. Edward Brown attributes this diminishment to a

loss of vigour as a result of changed conditions, [due] probably to limitation of exercise, but principally to the use of immature stock for breeding, as these birds do not reach maturity until they are three or four years of age. At the same time selections would be rather for medium than very large size, as the huge specimens are not suitable for ordinary farm purposes.

The first turkey was brought to Europe by a Spaniard in 1518. From Spain the turkey was taken to France and soon afterwards it was brought to Britain. Its common name of Turkey betrays the bewilderment the bird caused. At that time the general knowledge of the world's geography was slight and any place beyond the shores of Britain was foreign. People had the haziest notions of the definition of east and west and anything that arrived from abroad was automatically called Spanish or Turkish, so this bird, which was probably first carried to Britain by ships which called at Spain on their return from the Orient, was dubbed a Turkey. The French thought it came from India and gave it the name of Dindon or 'd'Indon'.

Initially turkeys were rare but slowly their numbers grew. André Simon comments that turkeys provided

the exception to the rule that it takes a long time to break down the prejudice of English people against anything new in the matter of food or drink. The *Turkey* was enthusiastically hailed and made welcome as soon as it appeared on the tables of the well-to-do in England. (*A Concise Encyclopaedia of Gastronomy* – 1956)

In 1541 Archbishop Cranmer, in order to conserve supplies, forbade the appearance of more than one dish of turkey cock at State Festivals and prohibited the eating of females altogether. These were needed for breeding and to boost their numbers more turkeys were brought in from abroad. Sir George Strickland introduced some to his home in Norfolk and, when he was granted arms by Edward VI in 1550, a turkey cock in 'his pride proper' was incorporated in his crest. Numbers of turkeys steadily increased and by 1555 there were enough for two males and two females to be served at a grand Law Dinner. By 1557 they were reasonably common and Tusser complains of the damage they did to the kitchen garden. Heresbach said they could not stand cold and wet and must be kept in a warm, dry place in winter and provided with ladders to reach perches which should be eight or ten feet from the ground, while Mascall considered their flesh, although delicate, too heavy and hard to digest and preferred eating peacock.

Others did not agree with Mascall's assessment for by the 1560s turkey

161

had become a Christmas dish and Tusser writing in 1573 says that this feast should be composed of

Beef, mutton, and port, shred pies of the best,
Pig, veal, goose, and capon, and turkey well drest

while an anonymous cynic stated that;

Christmas tidings of good cheer
To turkeys seldom sound sincere.

By the 17th century their meat was common enough for Hart to recommend in his *Diet of the Diseased* 'Turkies of a middle age and reasonably fat, are a good, wholesome, nourishing food, and little inferior to the best capon'. Turkeys were generally eaten 'either from the spit or in paste' and the meat was thought so delicious it compensated for the bird's inconvenient habits. Not only did turkeys require more housing than other poultry 'being an extream chill bird' but they needed more food to fill their 'large, stout, proud and maiesticall' frames. By the 18th century, everyone from farmer to cottager bred the birds and an industry had grown up round them in East Anglia.

These black Norfolk or Suffolk turkeys as they came to be known, were imposing birds with massive breasts and broad backs. Their long heads were covered by a reddish-pink skin 'wrinkled and formed into wart-like elevations' which projected over the beak in a drooping snood and descended in wattles down a neck held in a proud curve. They had hazel eyes and cushions of slatey black plumage which billowed around them and ended in a fan-shaped tail. D. H. Lawrence summons up some of their glory in his long poem *The Turkey-Cock* of which the following is an extract.

You ruffled black blossom,
You glossy dark wind.

Your sort of gorgeousness,
Dark and lustrous
And skinny repulsive
And poppy-glossy,
Is the gorgeousness that evokes my most puzzled admiration.

Your aboriginality
Deep, unexplained,
Like a Red Indian darkly unfinished and aloof,
Seems like the black and glossy seeds of countless centuries.

Armies of these magnificent birds were marched to English fairs and markets. Their feet were clad for the journey by being tied up in sacking and covered with a leather boot and at night they roosted in trees along the way. Daniel Defoe stated at the beginning of the 18th century that, 'These droves as they say, generally contain from three hundred to a thousand' and in autumn they filled the roads leading from East Anglia to London. The journey took at least a week and as many as '300 droves of turkeys . . . pass in one season over Stratford Bridge' while more went by other routes. In London the turkey's destination was the same as the chickens', the shops in Poultry. Later, because of the chaos caused by increasing numbers of birds and retailers, the markets were moved to other sites, including Leadenhall which is still the place where City gentlemen buy their Christmas turkey.

In the 19th century, at Christmas, 'From the low peasant to the lord, The Turkey smokes on every board' and it became fashionable to eat enormous birds. The Victorians had huge families and, perhaps as a way of demonstrating their new found industrial success, they liked to sit down to a table groaning under the weight of a gigantic bird. To increase their bulk turkeys were crammed with food by farmer's wives. One is described in the *Amicus Curiae* of 1847 as sitting

> with a leathern apron before her, with a bowl of warm milk, or some greasy water, taking the turkey out of the coop, onto her lap, forcing his mouth open, with her left hand, putting in the balls with the right, and stroking with her fingers the outside of the neck to make them descend.

On the continent, and especially in France, white birds were preferred because of their paler flesh which was considered more delicate. They plucked off the feathers before killing the bird in the belief that it tenderized the meat, a habit Cobbett denounced saying 'that the man that can do this, or order it to be done, ought to be skinned alive himself'.

The demand for large birds led to a search for new blood in America. Here vast birds were traditionally eaten at Thanksgiving and they specialized in producing males weighing twenty to thirty lbs at eight or nine months old. Because of continual infusions of vigorous wild blood American turkeys had remained stronger than English ones as Edward Brown in *Races of Domestic Poultry* explains.

> Scores of cases are recorded where a wild gobbler from the woods has taken possession of a flock of common turkeys, sometimes after first battling with and killing the domestic gobbler. The results of such a cross in almost every case have been so satisfactory that such meetings are much desired by turkey-raisers in those districts.

The Americans also crossed wild bronze turkeys with Mexican black turkeys and called the result the Black Bronze, later shortened to the Bronze. The credit of introducing these birds to Britain has been given, variously, to Lord Leicester of Norfolk, Lord Powis of Wales and Lord Derby of Lancashire; but, whoever it was, these brilliant copper-coloured birds with wing feathers edged in black made a great impact. 'The hens were good layers, making most faithful sitters and mothers'. Their flesh might lack the fine texture and taste of the Norfolk but their quantities of flesh met the criteria of the Victorians and the Bronze became established in Britain.

A cross between the Bronze, the black Norfolk and a grey turkey from southern Ireland, called the 'Bustard' with a 'wonderful quality of flesh . . . probably the finest in the world' produced the Cambridge Bronze. This bird was dowdy-looking and its colour varied from 'being grey, pied black and white, and rusty brown' but it had the advantage of a light frame onto which it laid quantities of 'soft flesh of great thickness'. As proof of the excellence of the Cambridge Bronze 'that county has for several years past supplied the Royal table on Christmas Day'.

Spurred on by these successes breeders sought new birds from America and other crosses. A union between the Bronze and a White turkey produced a buff-coloured bird and others with American breeds such as the Bourbon Red, black Narragansett and a Dutch bird called the White Holland made further hybrids.

The fattest turkey on record weighed 78 lbs 11¼ ozs but for commercial use strains which produce hens of 11½ lbs and cocks of 17 lbs at 18 weeks old are more popular. Strains free of the disease *mycoplasma melagridis* are also sought after as most turkeys are now raised under intensive conditions which means that this and other diseases are a problem. Their breeding has become increasingly specialized. Turkeys are usually hatched by smallholders who rear anything from 200 to 4000 birds and then sell them at one day old or at five weeks to farmers who fatten them for the table. The birds are crowded into houses but not, usually, onto wire or slatted floors as these can give them breast blisters and leg problems which affect the meat and the producer's profit. However turkeys do suffer the humiliation of debeaking and toe clipping to prevent any cannibalism, feather pulling, scratching and tearing that can occur because of over-crowding. Some even have their fatty beak snood clipped to prevent them getting head injuries if they fight. The birds are fed special high protein soya-based feed which increases their weight at such an unnatural rate that 5% die of heart attacks before they reach maturity.

Many turkeys are now bred so heavy they are unable to mate and the dismal sexual performance displayed by the modern super-fat cocks apparently dismays the hens to such an extent they have become uninterested in sex and their fertility has fallen. As usual science has produced a solution to its own manufactured failings and, in this case, it is artificial insemination. In dreadful mechanical fashion, two or three times a week

The tom is stimulated by stroking its abdomen and pushing the tail upwards and towards the head. The male copulatory organ enlarges and partially protrudes from the vent. By gripping the rear of the copulatory organ with the the thumb and forefinger from above and fully exposing the organ, the semen is then squeezed out with a short sliding downward movement. The males soon become trained and ejaculate easily when stimulated.' (*Raising Poultry the Modern Way* – Leonard S, Mercia, 1975)

The semen is collected in a glass funnel and then syringed into the female if, after manipulation, her oviduct can be exposed or protruded as this shows she is in laying condition.

The globe-shaped birds produced by these methods are eaten in ever-increasing amounts. At Christmas in 1981 it was predicted that 8·2 million oven-ready turkeys and 2·2 million 'traditional farm fresh' birds would be eaten and, over the rest of the year, more would be consumed. Wrapped up as turkey portions and turkey roasts the meat is becoming almost as commonplace as chicken and in the last twenty years the number slaughtered has risen from 3 million to 23 million and turkey rearing is described as a 'growth industry'. 73% of these birds are kept under intensive conditions and almost half the flocks are still in Norfolk with some other local units in Yorkshire, Aberdeenshire, East Lothian and Renfrewshire.

The turkey has come a long way from its Mexican tree roost. Human greed for its fine flesh, financial margins and a lack of space have made it descend to earth, demeaned its natural dignity, and suppressed its turkey-gobbling pomposity.

Conservation

The concept of conserving animals did not occur to man until the 19th century. Before this time man's relationship with animals was more balanced; it was made up of a mixture of 'love, exploitation and destruction' similar to that of the animals' own world. Primitive man had a kinship and respect for animals, first as hunter and then as farmer. He was dependent on them for most of his needs and admired them for their superior strength, speed, potency and heightened sense of smell, hearing and sight. Later this admiration became a kind of worship. Animals were endowed with a life force greater than man's own and gradually a dual-relationship and standard was established. This dual-relationship took many forms. From the admiration of animals as a superior force grew the idea of symbolism. In the ancient world certain animals were made totems of religion, and sacrificed to appease gods who committed strange and awful deeds. Yet at the same time man felt an intense love for animals which was reciprocated. The lamb was put under the protection of God and revered for its innocence and gentleness and Achilles' horses on learning that their charioteer was dead refused to move.

Through the centuries this double standard persisted. Certainly, as farming developed and hunting ceased to be a necessity and became, on the surface, something done for sport and social status, underneath it was also something done in response to an undefined, ancient instinct which lies deep in the psyche of animal and man making both take pleasure in violence. Kenneth Clark sums up this ambiguous attitude in his book *Animals and Men* (1977)

> We love animals, we watch them with delight, we study their habits with ever-increasing curiosity; and we destroy them. We have sacrificed them to the gods, we have killed them in arenas in order to enjoy a cruel excitement, we still hunt them and we slaughter them by the million out of greed.

It was the development of man's intellect and powers of speech that gave him the upper hand over animals. These enabled man to articulate his experiences, form ideas, exchange information and devise methods to circumvent the superior strength of certain animals. Once contained and given food and comfort the animals became quite tame and docile. Before domestication animals had been adapted to their natural environment by the survival of the fittest. After domestication man encouraged qualities which increasingly divided them from their original habitat and evolved artificial systems and breeding programmes to hasten the process.

All breeds of domestic livestock have been created by man through deliberate selection. The process culminated in the 18th century when

Robert Bakewell and his contemporaries solidified and classified these types into breeds. The theories they evolved gave man a hitherto undreamed-of manipulative power over the breeding of animals. Heady with their newly acquired skills many men imposed their aims on animals without stopping to think why, or for what purpose, certain characteristics had developed. They thought there could be nothing better than the New Leicester sheep and put its blood into almost every other one. From this moment on sheep genes began to be lost. Throughout the rest of the 18th century and in the 19th century individual characteristics that had evolved over centuries disappeared at an alarming rate as breeders 'improved' indigenous stock to fit the criteria of the new individual age.

David Low in his *Domesticated Animals of the British Isles* published in 1842 regretted what an infusion of Leicester blood had done to the Cheviot sheep, saying that

> there cannot be a question that, for general cultivation in the high and tempestuous countries to which the Cheviot breed is adapted, the race should be preserved in its natural purity. Every mixture of blood has been found to lessen that character of hardiness which is the distinguishing character of the race.

Low also bewailed the fact that the Romney Marsh sheep had become lighter and less hardy and that the 'Polled Irish Breed' of cow was dying out. Charles Darwin was another protestor, while in 1912 Lydekker in his seminal book on *Sheep* expressed the view that

> It is a matter for regret that these interesting old breeds of British sheep were allowed to die out without specimens, or at all events skeletons, being preserved in the national museum; but at the time of their disappearance little or no interest was displayed in the matter of preserving records of such vanishing types.

The relationship the average man had with animals underwent an enormous change in the 19th century. From being a nation of country dwellers Britain became predominantly one of townspeople. This change ended, for most people, a symbiotic relationship with animals. No longer was their care, feeding and breeding part of the daily life of almost every man whether farmer or cottager: living in towns and having to buy the food he had formerly produced himself distanced man from animals. The attitude of most townspeople altered dramatically; those who did not grow barbaric through lack of contact with animals became sentimentally attached to them. At the same time the farmer, compelled by a need for profit and an expanding population who required feeding, exploited his animals to their utmost production. In between townsman and farmer stood an ever-growing group of zoologists and naturalists who investigated animal behaviour to points formerly thought impossible. The combined effect of these three groups was to make some men adopt a new role towards animals as 'neither the worshipper nor the destroyer of animals, but as a kind of governess'.

Preservation groups began; in 1824 the Royal Society for the Prevention of Cruelty to Animals was founded, with the humanitarian William Wilberforce as one of its founder members, and the aim 'to promote kindness and to prevent or suppress cruelty to animals'; in 1891 the Royal Society of Bird Protection was formed; and in 1894 the National Trust to preserve land and houses.

Although these groups showed an awareness of the plight of nature they did nothing to prevent the destruction of breeds and, as the 20th century progressed, this continued. By the early years twenty-three breeds of domestic livestock had become extinct. The cattle breeds that went included the Alderney, Suffolk Dun, Sheeted Somerset, Castle-Martin and Caithness and the Irish Dun. Sheep losses included the Limestone, St Rona's Hill, Norfolk Horn, Roscommon and Welsh Rhiw. The pigs that disappeared were the Ulster White, Small White, Yorkshire Blue and White, Dorset Gold Tip, Lincolnshire Curly Coat and Cumberland. Horses comprised the Manx, Cushendale, Tiree, Long Mynd, Galloway and Goonhilly. Many of these animals could be of enormous value today. The Suffolk Dun was famous in the 18th century for its ability to produce quantities of milk even 'though subject to careless treatment, and supported on the most common kinds of food'; the Limestone sheep was a unique hill breed able, like the lowland Dorset Horn, to give birth at all times of year; the Lincolnshire Curly Coat was a great out-door pig with a robust constitution and a thick, weather-protecting, brillo-pad coating of hair. Some of the other breeds that disappeared were perhaps not so important but, even so, it can never be known if their own particular qualities might not have come in useful at a future date.

It has recently been recognized that, for instance, the Chillingham White Park cattle, which have existed in a semi-wild state since the park was created in 1270 have a unique set of unartificially-selected genes. And they have lean meat, like a wild animal, which could prove extremely useful to breed into over-fat cattle. The worker of the 18th and 19th centuries led an active life and wanted fat meat to provide energy but the desk-man of the 20th century has a sedentary existence and requires lean meat. Lean meat is at a premium today but there are few breeds able to supply it; two cows with this quality are the White Park and Longhorn and there are several breeds of sheep, among them the Hebridean, Manx Loghtan and Soay.

Many old breeds have become highly tuned to their environment. The Soay have lived for centuries on the bleak island of Soay and have adapted themselves to surviving on very little food in extreme weather conditions. The Exmoor pony has also evolved features which fit it for a harsh environment. It has an extra tooth, a seventh molar, which helps it masticate coarse herbage, a spacious nasal cavity which enables it to warm intakes of cold air, and a thick, wirey outer coat which sheds the rain and gives protection to the vital and sensitive parts of its body. The Galloway cow's coat consists of two types of hair, each electrically charged. The inner, protective coat is positive and the outer, water-proof one is negative. When the wind blows over the hair the charges strengthen, making the hairs cling closely together and keeping the Galloway warm and weather-proof.

These animals are often extremely economical eaters. Because of the scarcity of grazing in their home environment their bodies make efficient use of any food. The Kerry, Shetland and Northern Dairy Shorthorn cows, the Shetland and Hebridean sheep and the Gloucester Old Spot and Tamworth pig have never been molly-coddled by man and they can all scrape a living where others would starve. They actively search out their food by ranging far and wide. This characteristic could be useful if more margainal land has to be brought into production.

Primitive sheep have very soft wool. They grow two layers, an outer weather-proofed, hairy coat and an inner, soft, warm lining. In spring the inner layer falls away leaving a summer coating of hair. The warm layer grows again in autumn to help protect the animal from the winter cold. If you lose these breeds you lose wool that rivals the Cashmere in softness.

Many of these so-called minority breeds are also extremely fertile. The Teeswater, an ancient North Country hill breed, is noted for its prolificacy and will produce four and even five lambs. The Finnish Landrace is another fertile breed and its average litter size is three. This sheep was almost extinct but is now one of the most popular in Europe because of this quality.

Unimproved animals live for a long time. Modern breeding systems select for early maturity but in nature this is equated with a shorter productive life. The average milk-producing life of a modern dairy cow is four years but many of the older breeds will continue giving the liquid until over twenty. Modern sows seldom last beyond their third litter whereas breeds like the British Lop are still giving birth in their teens and even twenties.

Rationalizing to a few breeds means that you eliminate many of these useful characteristics. The genetic variety they represent could be crucial not only to the development of new breeds but also to the hybrids on which modern intensive systems thrive. Once lost these genes are irrecoverable as they cannot be created by man. An example of this can be seen in the attempts to resuscitate the Norfolk Horn sheep. In 1969 their numbers were down to six rams and fourteen ewes and these were horribly inbred. The males were cryptorchid and the females were infertile or produced lambs too weak to live. In an effort to revive the breed they were crossed with another similar sheep, the Wiltshire Horn, and with Suffolks, which were originally produced from the Norfolk. The result was a sheep that looked like a Norfolk Horn but could never be a true Norfolk Horn.

No matter how perfect the looks, like reproduction furniture or architecture, the intrinsic quality has gone. This is yet another argument for preserving the breeds. Another appeal for their protection could be made on the grounds used by people attempting to conserve old houses and monuments, that the animals represent part of our history.

Quite apart from their historical association many of the older breeds have fabulous looks. The admiration primitive man felt for the beauty and strength of animals has never disappeared. In more evolved societies animals have been the inspiration for great works of art and today thousands pay to look at them in Farm Parks.

The fact that many valuable breeds had already gone and others were in danger of extinction began to dawn on a few people in the 1960s. The Zoological Society of London, in an effort to help, established some units of endangered breeds at Whipsnade Zoo in 1964. But Whipsnade had hardly enough space for its wild animals let alone domestic ones and they were soon passed to the University of Reading. In 1968 the University also decided they did not have room but, at this point, the Royal Agricultural Society began to take an interest. It brought the remaining stock of Soay, Cotswold and Portland sheep to its farms at Stoneleigh in Warwickshire and, in 1971, set up a working party to review the situation and instigate a programme for the preservation of rare breeds. This led in 1973 to the formation of the Rare Breeds Survival Trust and its members immediately carried out a survey of the current state of British breeds. They discovered that forty-five, or about half the animal breeds were in imminent danger of extinction and that virtually all pure breeds of large fowl, most breeds of duck and several breeds of turkey were very rare.

This concern in Britain coincided with similar feelings in Europe. In 1966 the United Nations, through its Food and Agriculture Organization, instituted a series of meetings on breed preservation and, in 1968, the Biosphere Conference in Paris recommended that domestic animals should be preserved in order that 'the rich variety in their genes will not be forever lost because of the present tendencies in agriculture and animal husbandry to concentrate on a limited and highly selected array of strains'. Then, in 1972, the Stockholm Conference on the Environment suggested that a catalogue of endangered domestic animals be compiled. Three years later, using the material they had gathered, the FAO set up a Pilot Study on the Conservation of Animal Genetic Resources. This research was later extended to cover Third World countries where some totally unsuitable western breeds were being imported out of ignorance. These were gradually supplanting the indigenous animals which were better suited to the rigours of tropical life and, in many cases, were resistant to endemic diseases. The Kuri cattle, an ungainly breed which produces a moderate amount of milk and excellent beef, is nevertheless admirably suited to its life on the shores of Lake Chad. Here the animals have to subsist on coarse vegetation and swim from island to island to reach new pasture. They swim with their heads tilted back, lying their large, lyre-shaped horns on the surface of the water which increases their buoyancy and manouverability.

The British Rare Breeds Survival Trust (whose headquarters are now at the National Agricultural College, near Stoneleigh in Warwickshire) in an effort to ascertain which breeds should be strengthened and preserved devised an acceptance procedure. Each animal had to breed true to type, have existed for seventy-five years with other breeds contributing less than twenty per cent of their make-up and must have an official herd book going back six generations. The RBST will help breeds of cattle when their numbers fall below 740, sheep 1,500 and pigs 150, and give priority to those decreasing significantly or which are found in fewer than four units more than fifty miles apart. They assess each breed for its genetic value, purity, distinctiveness, antiquity, historical importance and current commercial qualities and then take steps towards its preservation by encouraging breed societies. The RBST have also established a semen bank, investigated the long-term storage of embryos and computerized breeding lines to help strengthen blood lines. Their work has ensured that there are now no known breeds in danger of extinction and the awareness of the general public has been enormously increased.

The preservation of the minority breeds has been vindicated by recent events in agriculture. The idea that became current in the 1950s, that only a few breeds were needed, is now discredited. The earlier idea was based on a glut of farm produce and a glut of oil, which created cheap energy and the cheap by-products of fertilizer and concentrated feeds. It was thought that what was required were a few, large-yielding, food-guzzling animals to meet the needs of the system. The Friesian cow, it was thought, admirably suited the needs of the dairy industry as each animal produced over 1000 gallons of milk a year; the Charolais and other large-framed exotics from the continent fitted the beef bill; the Border Leicester, Suffolk and Blackface sheep the wool and mutton market; the Landrace and various hybrids the pork demands; and hybrid birds achieved the aims of the poultry industry. The Norwegians even went as far as instigating a policy designed to eliminate all their breeds of cow but one, the Norsk Rodtfe.

Then the effect of all this specialization and rationalization became apparent. It was seen that the body of the Landrace pig had been grown too long for its legs to support and that its meat was so lean there was no fat to hold the tissues together and it separated when sliced at high speeds.

The enthusiasm for the creamy-coloured Charolais was tempered by the discovery that it has difficulty giving birth and the calves are frequently born dead. The problem is a direct result of a breeding policy that has concentrated on quick growth to high weights. Unborn calves are of such advanced muscular development they are unable to slip easily from the womb.

Faults due to over-breeding can only be corrected by infusing new, vigorous blood and the main source of this is, increasingly, the older, unimproved breeds. These self-sufficient breeds may also be required if the rules of factory farming are changed. Its whole ethos is under scrutiny for several reasons. Recently it was established that feeding high-energy diets to animals deprived of exercise results in the accumulation in the body of a high ratio of fat to protein. This amounts to approximately three to one whereas free-ranging animals, able to select their own feed, have a fat ratio of one to three. Intensively produced meat is also of low quality and lacks the essential fatty acids needed by humans for cell growth and body maintenance.

The cost of over-crowding these animals and producing their semi-poisonous meat rises daily. Their houses become increasingly expensive and the oil needed to provide the energy to regulate their temperature, humidity and light not only costs more but gets scarcer. The cost of drugs required to solve the problems of intensive farming is now astronomical and a leading pig producer recently admitted to spending more on drugs than on feed.

Feeding these 'high-growth' animals now takes up about nine-tenths of our farm land and requires substantial imports of feed stuffs. As land gets scarcer, due to the spread of towns and motor-ways, and artificial feed gets more expensive because it relies on oil, fewer, more economical animals could be required. Fewer animals might also be dictated by the consumer. The high cost of producing food is reflected in retail prices, making people turn to cheaper cereal and pulse-based foods. When they buy meat, they increasingly want good meat. The old idea of growth for growth's sake took no account of the quality of the food and it may be that this situation will change in the future. The long-term solution could be to have smaller groups of animals, tended by greater numbers of stockmen, on more farms. This would result in a better product, better standards of animal welfare, increased employment and fewer problems with anti-biotic drug resistance.

No one knows what the future holds but since change has been the invariable rule of history and there is no reason to think this pattern will alter, then allowances should be made for the unexpected. Already the alternatives that do exist have been called into use. The large frames of the Lincoln and Leicester sheep have been found to fit modern cooking requirements. A few years ago people liked small joints but now they prefer pre-packaged lumps of meat to make into casseroles and for this the larger sheep are well suited. The Whitefaced Woodland and the Lleyn sheep have recently been proved as profitable as modern cross-breds. The Jacob sheep was almost extinct in 1968 but is now a viable part of the British sheep industry. The Wensleydale recently came top of some meat trials and sophisticated tastes have discovered the sweet meat of the Black Welsh Mountain and smoked Portland lamb. The Longhorn cow won the top prize for beef at the Royal Show in 1981. The Red Poll is well adapted to use mixed diets and will produce 75% as much milk as the Friesian. The Cornish Fowl had fallen into disuse until it was recognized it could provide growth genes for broiler chickens. All these instances can be multiplied and increase the arguments for keeping all the options open.

The rights of animals also deserve to be recognized. Factory farmers metamorphose them into food machines and claim that the provision of sufficient food, water and warmth is enough for their well-being while overlooking their mental and physical frustration. It is increasingly being asked whether it is even reasonable to produce quantities of often unnecessary food at the expense of such large-scale suffering. Breeders who have taken steps to improve the welfare of veal calves by releasing them from crates and keeping them loose in houses and allowing them free access to milk, water and roughage, and pig producers who do not confine pregnant sows in narrow spaces without bedding, actually find production levels increase.

A fair deal is required for the 400 million farm animals slaughtered annually in Britain. During his life-time the average Briton eats an estimated 7 bullocks or heifers, 1 cow, 36 sheep, 36 pigs and 550 poultry but most people are divorced from the mechanism of animal production and have an ambiguous approach to the industry. The 'mutilated travesties bred for man's convenience rather than their own welfare' are sold by advertisements which show idyllic country scenes, and many farmers dismiss concern over their welfare as the sentimentality of the townsman. But by debasing animals, man debases himself. A feeling of kinship for a pathetic, battered, battery hen or a slippery, pink, attenuated, intensively-produced pig may be a difficult relationship to establish, but a feeling of kinship for the bright personalities personified by many of the unspoiled breeds is quite possible. It is strange that in our modern, social state we go to extremes to help the poor and spend millions on pet foods but have only just begun to pay attention to the plight of those animals on which the life of *our* species has, over thousands of years, depended.

Select Bibliography

Alderson, Lawrence: *The Chance to Survive – Rare Breeds in a Changing World*, Cameron & Tayleur in association with David & Charles, Newton Abbot, 1978
The Observer's Book of Farm Animals, Frederick Warne & Co. Ltd., 1976
Ark Magazine, the monthly journal of The Rare Breeds Survival Trust, 1974–82
Bonser, K. J.: *The Drovers – Who they were and how they went: An Epic of the English Countryside*, Macmillan, 1970
Brown, Edward, F.L.S.: *Races of Domestic Poultry*, Edward Arnold, London, 1906
Brown, J. T., F.Z.S.: editor *The Encyclopaedia of Poultry*, Walter Southwood & Co. Ltd., London
Burton, S. H.: *Exmoor*, Robert Hale & Co., 1974
Cameron Sillar, Frederick and Meyler, Ruth Mary: *The Symbolic Pig – An Anthology of Pigs in Literature and Art*, Oliver & Boyd, Edinburgh & London, 1961
Carsdale, G. S.: *Animals and Man*, Hutchinson, London, 1952
Chivers, Keith: *The Shire Horse, A History of the Breed, the Society and the Men*, Futura Publications, 1978
Clare, John: *The Shepherd's Calendar*, ed: Eric Robinson and Geoffrey Summerfield, Oxford University Press, 1973
Clarke, Kenneth: *Animals and Men*, Thames & Hudson, 1977
Clutton-Brock, Juliet: *Domesticated Animals from early times*, Heinemann/British Museum (Natural History), 1981
Cobbett, William: *Cottage Economy*, 1st ed: 1822, Oxford University Paperback, 1979
Coppock, J. T.: *An Agricultural Atlas of Scotland*, John Donald Publishers Ltd., Edinburgh, 1976
David, Elizabeth: *French Provincial Cooking*, Michael Joseph, London, 1969
Dent, A. A. and Machin Goodall, Daphne: *The Foals of Epona: A history of British Ponies from the Bronze Age to Yesterday*, Galley Press, London, 1962
Druid, The: *Saddle and Sirloin*, Frederick Warne & Co.
Easom Smith, H.: *Modern Poultry Development: A History of Domestic Poultry Keeping*, Spur Publications Co., Hill Brow, Liss, Hampshire, 1976
Elton, Charles: *The Ecology of Animals*, Chapman & Hall, (first published in 1933), 1966 edition
Ernle, Lord: *English Farming Past and Present*, Longmans, Green & Co., 1936

Ewart Evans, George: *The Horse in the Furrow*, Faber & Faber, 1960

Fiennes, Celia: *The Journeys of Celia Fiennes:* edited and with an introduction by Christopher Morris, Cresset Press, London, 1947

Firbank, Thomas: *I Bought a Mountain*, George G. Harrap & Co. Ltd., London, 1940

Fraser, Allan: *The Bull*, Osprey Publishing Ltd., Reading, Berkshire, 1972

Fraser, Andrew F.: *Farm Animal Behaviour*, Barriere Tindall, London, 1974

Gilbey, Sir Walter: *Concise History of the Shire Horse*, 1889 reissued by Spur Publications Co., Hill Brow, Liss, Hampshire, 1976
Farm Stock of Old (1st ed: orginally published as Farm Stock 100 Years ago, 1910); reissued by Spur Publications in 1976

Glover, Janet R.: *The Story of Scotland*, Faber & Faber, 1977

Gosset, Adelaide L. J.: *Shepherds of Britain – Scenes from Shepherd Life Past and Present*, Constable & Co., 1911

Hammond, John MA, Dsc (Iowa) FRS: *Farm Animals, The Breeding, Growth and Inheritance*, Edward Arnold & Co., London, 1941

Harnett, Cynthia: *The Woolpack*, Penguin, 1965

Hartley, Dorothy: *Food in England*, Macdonald and Jane's, London, 1975

Hartley, Marie and Ingilby, John: *The Old Hand-Knitters of the Dales*, Reprint of original 1951 edition by The Dalesman Publishing Company, Clapham, Lancaster, 1978

Holmes Pegler, H. S.: *The Book of the Goat*, 5th ed. The Bazaar Exchange and Mart Office, London, 1917

Hoskins, W. G.: *The Making of the English Landscape*, Book Club Associates, London, 1977
History from the Farm, Faber & Faber, 1970
Provincial England, Macmillan, 1963

Hudson, W. H.: *A Shepherd's Life*, Methuens Modern Classics, 1933

Hulme, Susan: *The Book of the Pig*, Spur Publications, Saiga Publishing Co. Ltd., 1 Royal Parade, Hindhead, Surrey, 1979

Irwin, John: *The Kashmir Shawl*, Victoria & Albert Museum, HMSO, 1973

Isaac, Peter: *The Farmyard Companion*, compiled by Peter Isaac, Jill Norman & Hobhouse Ltd., 1981

Low, David: *On the Domesticated Animals of the British Isles* 4th ed., 1846

Mackenzie, David: *Goat Husbandry*, Faber & Faber, 1957

Maclean, Charles: *Island on the Edge of the World: The Story of St. Kilda*, Cannongate, Edinburgh, 1977

Moore, Ian: *Grass and Grasslands*, Collins, 1966

National Sheep Association: *British Sheep*, 1976

Nicolson, James R.: *Traditional Life in Shetland*, Robert Hale, London, 1978

Pawson, H. Cecil: *Robert Bakewell, Pioneer Livestock Breeder*, London, 1957

Power, Eileen: *The Wool Trade in English Mediaeval History*, Oxford University Press, 1955

Scott Watson, James A., and Hobbs, Mary Elliot: *Great Farmers*, Faber & Faber, 1951

Seebohm, M. E.: *The Evolution of the English Farm*, revised 2nd ed., George Allen & Unwin Ltd., London, 1952

Simon, André L.: *Guide to Good Food and Wines: A Concise Encyclopaedia of Gastronomy*, Collins, 1956

Stewart Collis, John: *The Worm Forgives the Plough*, Penguin, 1976

Taylor, Christopher: *The Making of the English Landscape: Dorset*, Hodder & Stoughton, 1970

Trevelyan, G. M.: *English Social History*, Longmans, 1973
A Shortened History of England, Penguin, 1974

Trow-Smith, Robert: *A History of British Livestock Husbandry* Vol I: to 1700; Vol II 1700–1900, Routledge and Kegan Paul, 1957
Life from the Land: The Growth of Farming in Western Europe, Longmans, 1967

Toulson, Shirley and Godwin, Fay: *The Drover's Roads of Wales*, Wildwood House Ltd., London, 1977

Whitehead, G. Kenneth: *The Ancient White Cattle of Britain and Their Descendants*, Faber & Faber, 1953
The Wild Goats of Great Britain and Ireland, David and Charles, Newton Abbot, 1972

Whitlock, Ralph: *The Land First*, Museum Press, London, 1954
Rare Breeds, Prism Press, Stable Ct., Chalmington, Dorchester, Dorset, 1980

Williams, David: *A History of Modern Wales*, John Murray, London, 1950

Williamson, Henry: *The Story of a Norfolk Farm*, Faber & Faber, 1941

Orwell, George: *Animal Farm*, Penguin Books, 1977

Youatt, William: *Cattle, Their Breeds, Management and Diseases*, Baldwin and Craddock MDCCCXXXIV
The Horse, Longman Green, London, 1861
The Pig – A Treatise on the Breeds, Management, Feeding and Medical Treatment of Swine, Craddock and Co., London, 1847
Sheep, The Breeds, Management and Diseases, Robert Baldwin, 1937

Zeuner, F. E.: *A History of Domesticated Animals*, Hutchinson, London, 1963

Index